Energy *Survival* Guide

Jo Hermans

ENERGY SURVIVAL GUIDE

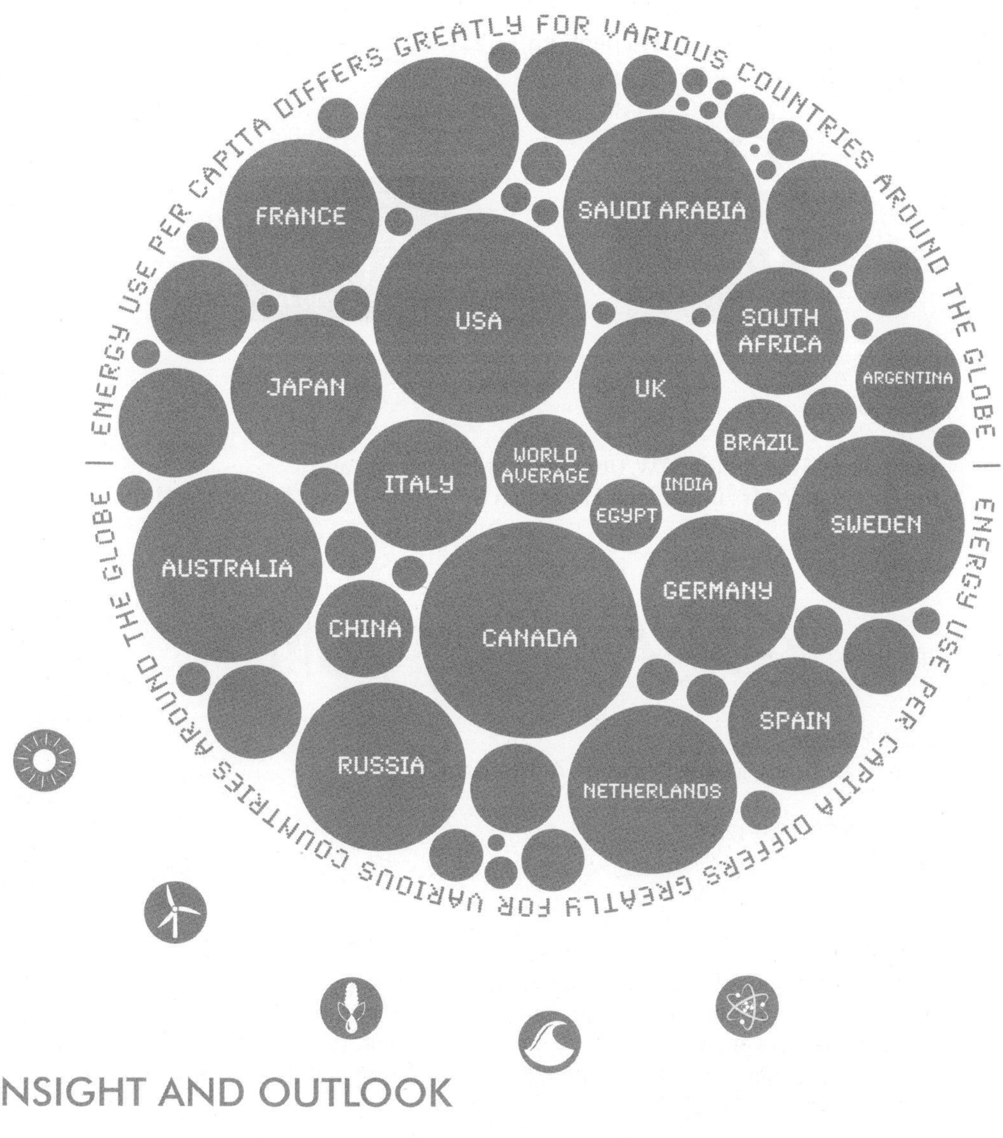

INSIGHT AND OUTLOOK

Published by

BetaText, Prins Bernhardlaan 10, 1862 EM Bergen NH, the Netherlands
www.betatext.nl

Leiden University Press, p/o Rapenburg 73, 2311 GJ Leiden, the Netherlands
www.lup.nl

In co-operation with

Technology Foundation STW, Utrecht, the Netherlands, and
the Foundation for Fundamental Research on Matter (FOM), Utrecht, the Netherlands

Graphic design
Michael de Groot, Misign Ontwerp, Amsterdam

ISBN 978 90 8728 123 6
NUR 961

To my grandsons David and Thomas
and their little friends around the globe,
who are the heirs to our successes and failures
in making the world sustainable

Part 1 | Insight

Part 2 | Outlook

Useful reference data

Conversion Factors

In the field of energy, exotic units are frequently used. Here are some conversion factors for energy (into joules, J) and for power (into watts, W).

Energy

1 calorie (cal)	=	4.187	J
1 kilocalorie (kCal, or even just Cal, very confusing)	=	4.187	kJ
1 kilowatt-hour (kWh)	=	3.6	MJ
1 British Thermal Unit (BTU)	=	1.055	kJ
1 ton of coal equivalent	=	29.3	GJ
1 ton oil equivalent (toe)	=	41.9	GJ
1 million barrels (159 litres each)	≈	5.7	PJ
1 electronvolt (eV) (used for atomic particles)	=	1.6×10^{-19}	J

Power (energy per unit of time)

1 kWh/y or kWh/a (kilowatt-hours per year)	=	0.114	W
1 J/y or J/a (joules per year)	=	3.171×10^{-8}	W
1 mbd (millions of barrels per day)	≈	6.6×10^{10}	W
1 BTU/h (British Thermal Units per hour)	=	0.293	W
1 BTU/a or BTU/y (British Thermal Units per year)	=	3.345×10^{-5}	W

Heats of Combustion

Carbohydrates, proteins			15	MJ/kg
Fat, oil			40	MJ/kg
Petroleum (standard)			41.9	MJ/kg
Petrol	35 MJ/litre	or	48	MJ/kg
Natural gas (1 bar; depending on origin):				
higher*			33–41	MJ/m³
lower*			30–37	MJ/m³
Hydrogen (0°C, 1 bar):				
higher*	142 MJ/kg	or	12.8	MJ/m³
lower*	120 MJ/kg	or	10.8	MJ/m³
Coal (standard)			29.3	MJ/kg
Wood (dry)			20	MJ/kg
Natural uranium: approximate yields at present**				
Standard LW-reactor, 'once through'			500	GJ/kg
Standard LW-reactor, incl. reprocessing			670	GJ/kg
Breeder reactor			34,000	GJ/kg

* The 'higher heating value' includes the heat of condensation of the water vapour formed; the 'lower heating value' does not.

** The fission energy of uranium-235 equals 80,000 GJ/kg.

Rule of thumb: 100 W ≈ human being ≈ candle flame ≈ ¼ litre oil/day

Useful reference data

Prefixes

kilo	k	10^3	(thousand)
mega	M	10^6	(million)
giga	G	10^9	(billion, in UK and USA)
tera	T	10^{12}	(trillion, in UK and USA)
peta	P	10^{15}	
exa	E	10^{18}	
milli	m	10^{-3}	(thousandth)
micro	μ	10^{-6}	(millionth)
nano	n	10^{-9}	
pico	p	10^{-12}	
femto	f	10^{-15}	
atto	a	10^{-18}	

Air and Water

Mean atmospheric pressure: 1.013 bar (1 bar = 10^5 pascal)
Density of air at 1 bar, 20°C: 1.2 kg/m^3
Heating air at atmospheric pressure by 1°C takes 1.2 kJ/m^3 or 1 kJ/kg
Heating water by 1°C takes 4.2 kJ/litre
Evaporating water takes 2.3 MJ/litre

Petrol, Diesel, CO_2-Emission

1 litre of petrol has a mass of approximately 0.75 kg
1 litre of diesel oil has a mass of approximately 0.85 kg
The emission per kg oil or petrol burned is 3 kg CO_2
The emission per m^3 of natural gas burned is 2 kg CO_2

Earth and Sun

Earth: circumference = 40,000 km; diameter = 12,730 km
Sun: diameter = 1.4 million km; apparent diameter in degrees = 0.5 degree
Average distance sun–earth = 150 million km
Mass of the sun = 2.0×10^{30} kg
Temperature on the sun's surface ≈ 5,750 kelvin or ≈ 5,500°C
Energy flow emitted by sun = 3.8×10^{26} watt
Energy flow emitted by sun in terms of mass loss = 4×10^9 kg *per second*

Solar Energy Flux

Just beyond the atmosphere (perpendicular to sun rays): 1,350 W/m^2
On the surface of the earth (perpendicular to sun rays): 1,000 W/m^2
Year-around day/night average, on horizontal surface in N-W Europe: 110 W/m^2
(per year this amounts to 3.5 GJ, or 90 litres of oil per m^2, a 9 cm thick layer)

Energy use (in kW/capita, on a continuous basis)

World average	2.4 kW
France, Germany, Japan	5.5 kW
USA, Canada	10 kW

Preface

This book serves as a true *Survival Guide* in two senses.

First, it is a guide for all of us facing rising energy costs. Chapters 1 to 7 provide *insight* into the concept of energy and the role it plays in our daily life. A little insight yields a great deal of understanding of energy and its costs.

Second, this book serves as a reliable source of information for anyone who wants to have a well-balanced opinion about the energy future of our planet. To this end, chapters 8 to 12 sketch the *outlook* into our energy future. Alternatives for the post-fossil-fuel era are reviewed. Their potential and their pros and cons are assessed.

Talking about energy without using numbers is futile. It is essential to look at the quantitative aspects. This does not mean that things get complicated. Mostly, applying a simple rule of thumb with some elementary arithmetic will do. Sometimes it even suffices to use a simple – yet powerful – tool: the law of conservation of energy. Useful reference data frequently needed are given on pages 8 and 9. Since insight and easy reference are given priority over high precision, rounded-off numbers are frequently used.

For readers who just want to have an easy overview of a certain topic, it may suffice to read the conclusions at the end of the relevant chapter. Those preferring extensive information are invited to read the complete text. Readers who wish to understand things in more detail will find turquoise 'boxes' (for example, **Counting flames** on p. 27). And for those who enjoy doing a bit of science there are 'boxes' indicated by a '**cartoon professor**'. These contain further information which requires a minimum of secondary-school physics to understand.

When discussing practical cases – domestic energy use, for example – it is often necessary to be specific and take a certain country. If energy use in an average industrialised country is considered, the Netherlands is sometimes used as a case study. For many other countries having comparable life styles, the data are rather similar on a per-capita basis.

The author hopes that this book will be useful for those who wish to make wise choices in their private life, and that it will contribute to a responsible future energy policy for all.

Jo Hermans
Leiden, November 2011

1 | Energy: A Truly Special Case

Our standard of living in the developed world is intimately linked to the easy availability of energy. All of the convenient appliances that we are used to – our fully automatic heating systems at home, our stereos, TV, dishwashers, refrigerators, freezers, computers, laptops, mobile phones, GPS – all of these appliances would be impossible without the silent flow of energy into our homes, through the electrical cables and the natural-gas or oil pipes. When we leave our home, we can get around quickly and efficiently by car, train, bus, tram, subway and aeroplane. Gone are the days of a slow ship that took a full week to transport us from Europe to the US, for example. The availability of cheap energy has made our lives pleasant and saves us a lot of time. Without inexpensive energy these amenities would simply be unavailable in modern life.

A striking example of the importance of energy is the Dutch 'Golden Age', the 17th century. Why was Holland – a relatively small country – so powerful and prosperous? It is not because the Dutch were so much stronger or smarter than their neighbours. By and large, it was the result of cheap and easily available energy. First of all, Dutch soil was composed of a lot of peat, which is a much handier fuel than wood. Digging peat and transporting it by sailing ship was immensely more efficient than cutting wood and transporting it by a horse-drawn cart along the road, which was how people in the neighbouring countries were doing it. Moreover, the Dutch live in a flat and windy country, so they built windmills.

The Dutch 'Golden Age' born out of peat and windmills

These produced the mechanical energy needed for sawing wood, milling wheat and pumping water. Just imagine how you would do that without this kind of machinery. Horse and man power was the only alternative, and that is precisely what neighbouring countries were using. In addition, we should realise that those horses had to eat, so additional farmers were needed to produce food for the working men and for the animals.

It turns out that using these efficient forms of energy saved a total of about 165,000 people from having to engage in useless labour. This number, by the way, roughly corresponds to the total urban population of Holland at the time. Those people were now free to do other, more useful things, like engage in the arts and sciences. So we may safely conclude that the most important root of the Golden Age was the availability of easy energy. The Golden Age was 'born out of peat and windmills' [1].

The 'peat period' was followed by the coal period. And then came the age of *really* convenient energy: petroleum and natural gas. Meanwhile, the invention of the steam engine brought us convenient *mechanical* energy. This made the industrial revolution possible. Energy became even easier and cheaper. The industrialised world was born. Not out of peat, but out of coal, petroleum and natural gas.

Exactly how cheap our energy has become can be illustrated by presenting just a few figures. Take, for example, our domestic electricity consumption. In a country like the Netherlands, the average household consumes about 400 watts on a continuous basis, averaged over day and night. This is the equivalent of having 4 lights of 100 W on constantly. The electricity is provided simply by connecting our appliances to a wall socket, and this is how our refrigerator, washing machine, dishwasher, TV and stereo run and how we charge the batteries for our laptop, cell phone and iPod. What if we had to generate those 400 W with our bodies, for example by using an exercise bicycle equipped with a generator? We could never generate enough energy on our own. We would need 4 people pedalling continuously, as will be pointed out later on. And if we assume an 8-hour work day, we would need 12 people. This means that we have a dozen imaginary servants per household employed just to supply us with the electricity we use. And what is our current utility bill? It is probably less than €3 per day. This sounds like a bargain compared to the 'servants scenario'.

And now we're faced with the prospect of the depletion of fossil fuels. But isn't this just like any other natural resource that will eventually run out and for which we have to find a substitute or opt for recycling? This is not the case. Energy is fundamentally different from 'ordinary' resources. Take copper, for example. If we run out of copper, we can find other metals such as aluminium that would serve as a good conductor and that could replace copper in electrical cables.

1. This has been well documented, and accurate figures were established using old archives, see, for example, 'De Gouden Eeuw uit turf geboren' (The Golden Age born out of peat) by J.W. de Zeeuw, Spiegel Historiael, December 1979, p. 686–692, and 'Energiemarkten en energiehandel in Holland in de late Middeleeuwen' (Energy markets and energy trade in the late Middle Ages) by C. Cornelisse; ed. Verloren, 2008).

If we run out of steel, we can often replace that with new strong synthetic materials. And wherever necessary, we can always return to the junkyard or dump, and recycle our discarded things. After all, these materials never disappear from the surface of the earth, so they can always be recovered.

Unfortunately, this is not the case with fossil fuels. Once we have consumed them, they are gone. Their heat is radiated into space. It will require millions of years for nature to supply us with a new load of fossil fuels.

Many shortages can be solved as long as we have energy available. What about a shortage of water? Just build a desalination plant to convert sea water into drinking water. Or, if worse comes to worst, why not just go to the Arctic and tow an iceberg to Europe or Africa? These are actually feasible scenarios, provided we have enough energy. But if we don't have the energy, we are stuck. Thus, energy is indeed a unique case.

2 | Energy: What Exactly Is It?

Energy is a difficult concept to comprehend (see box p. 16). It comes in various forms: kinetic energy, potential energy, mechanical energy, electrical energy, chemical energy, and heat. These can all be transformed into one another. All of these forms and transitions are, however, governed by one firm law: the law of conservation of energy. This law states that the total amount of energy remains constant.
The end product is almost always heat. Take a bouncing ball. When it is up in the air, it has potential (in this case: *gravitational*) energy. Just before it hits the ground, all of that potential energy has been converted into *kinetic* energy. Then it gets squashed like a spring, and most of its energy is converted into potential (in this case: elastic) energy. Only a small portion of it becomes heat at the first bounce. Finally, when the ball stops bouncing, ALL of its energy has been converted into *heat*.

This also happens in other situations. Take our stereo, for example. It draws its energy from a wall socket. A substantial amount of energy is converted into heat in the amplifier, which we notice because it warms up a bit. The energy that goes into the speakers is also partly converted into heat. The remainder is released as sound. Eventually, even *that* part will be absorbed into our homes, and also converted into heat. So, in the end, *all* electrical energy drawn from an electrical socket is converted into heat.

The reverse process – converting heat into other forms of energy – is more difficult. If we want to convert heat into mechanical energy (as a steam engine or car engine does) only some of the heat is converted into mechanical energy. The rest will be released as waste heat.
In fact, there is a law of physics that puts an upper limit to the efficiency that this conversion process can have. This limit is completely determined by the temperature range of the process. To be precise, the maximum efficiency possible for this cyclic process is $1 - T_{low}/T_{high}$, the so-called Carnot efficiency. Here the temperatures T are in kelvin, *i.e.*, they are calculated on the 'absolute temperature scale', or 273 degrees + the Celsius temperature. T_{low} and T_{high} refer to the lowest temperature used in the cycle (usually the temperature of the cooling water) and the highest value (the temperature of the steam if we are discussing a steam engine). Increased efficiency requires the ratio T_{low}/T_{high} to be small. But the lower temperature is usually the ambient temperature (*i.e.*, 300 K); much lower would be impractical. The highest value is limited by the properties of the equipment used, which simply breaks down if the temperature is too high. Thus, it comes as no surprise that a petrol engine has an efficiency that is far below 100% (in fact, it is only about 25%).

Once we have mechanical energy, we can use a generator to convert it into electrical energy at a much higher efficiency, even 100% in theory.

The same holds true for the reverse process: converting electrical energy into mechanical energy, using an electric motor. There is no law of physics that prohibits a conversion efficiency of 100% in this case either. In practice, the efficiency will be somewhat

Energy, Work, Heat

If I engage in physical labour by hitting a squash ball, the ball acquires energy. It can, in turn, perform work by using that energy. It can break a window, for example, compress a spring or hit a second ball. Work and energy are therefore interchangeable, so they have the same unit: joule (rhymes with cool). Its symbol is J. This unit fits neatly into the international metric system of units, along with metre, kilogram and second. The joule as unit of energy equals the unit of force (the newton, N) times the unit of distance (metre, m). In other words: 1 J = 1 N m.
Eventually, work and energy are converted into heat: the squash ball will become noticeably warmer, after a few minutes of play. So the unit of heat is also the joule. The former and old-fashioned calorie equals approximately 4.2 joules. Consequently, a kcal (sometimes written as Calorie) equals 4.2 kJ.

Power

The rate at which work is done, energy is used, or heat is produced is called power. It is expressed in joules per second and has a special name: watt. Its symbol is W. In other words: 1 W = 1 J/s.
Conversely we have 1 joule = 1 watt for 1 second (a watt-second). This is analogous to a man-hour, or the work done by one man in one hour.
A familiar unit is the one used on our electricity bills: the kilowatt-hour (or kWh). This is calculated as follows: the rate at which the electric meter in our home turns is proportional to the number of watts we have 'on' in our home. Suppose that is 1,000 W (a vacuum cleaner turned on, for example). If we leave it on for an hour, it will consume 1,000 watt-hours, or 1 kWh. That is the unit that appears on our electricity bill. We can convert this into the standard energy unit (J) by noting that 1 kWh equals 1,000 W for a period of 3,600 s, or 3,600,000 Ws, or 3.6 million joules, or 3.6 MJ (megajoules).

The same goes for heat: an ordinary wooden match, for example, has a heat of combustion of about 2,000 joules. When it burns for 20 seconds, it uses 100 J/s, so it supplies 100 J/s of heat, or 100 W.

People sometimes talk about 'watts per second' or 'kilowatts per hour' as a unit of power. This is nonsense, of course, since 'watt' is already measured per unit of time. It is like noting the speed of a boat in 'knots per hour'; nonsense because a knot already *is* a mile per hour.

Energy and Power

lower, since it may not be economical to produce motors or generators with such a high degree of perfection. In addition, there is always some friction. A cheap generator may therefore be much less efficient, which you can tell from the amount of heat it releases. An efficiency of 90% is a reasonable expectation. Even so, electrical and mechanical energy can be converted into one another at an efficiency of up to 99% in practical cases. This is the case, for example, with the electric motor used in the 'Nuna', the solar car that has won the World Solar Challenge in Australia four times in a row in 2001, 2003, 2005 and 2007.

Conversion Processes and their Approximate Efficiency

Conversion process		Efficiency R (%)
Mechanical energy ↔ Electricity		
Generator, dynamo		80–99
Electric motor		80–99
Heat → Mechanical energy: $R < 1 - T_{low}/T_{high}$		
Steam turbine		40–60
Diesel engine		30–35
Petrol engine		20–25
Stirling engine		30
Heat → Electricity: $R < (1 - T_{low}/T_{high}) \times R_{mech \rightarrow el}$		
STEG*	maximum around	58
Power plant	average	40
At home, from the wall socket		35
Mechanical energy → Heat		
Via friction		100
Using a heat pump **		300–500
Heat → Heat		
Gas stove in the kitchen		50
Electric hot plate, induction plate: 35% × 70% =		25
Central heating unit (gas, 'higher heating value')		95
Solar radiation → Heat (solar panels)		20–70
Solar radiation → Electricity:		
Commercial silicon cells		12–17
Multilayer cells, in laboratory (maximum)		43
Solar radiation → Biomass (photosynthesis)		0.2–5
Food → Mechanical energy (muscles)		20–25

* *Steam and Gas: A 'gas turbine' here is powered directly by combustion of natural gas. Moreover, the hot exhaust gases generate steam which powers a steam turbine.*

** *The heat pump is an exceptional case, see page 28.*

Stirring in a pan

It may come as a surprise to most of us. Take a pan of soup, or – better yet – porridge or something else that takes a lot of energy to stir. When stirring it, where does the energy we expend go? The law of conservation of energy tells us that it must go somewhere. If we think about it, we find that there is no escape: all of the energy goes into heating up the food. And yet, we barely notice any increase in temperature.

The message here is that heating things up requires a lot of energy. Most people tend to think that energy is used primarily when lights are switched on, or things move (a fan, an electric razor, etc). But the law of conservation of energy tells us something else. If we want to find out where energy is used, we should neither think of motion nor light, as they do not offer a reliable yardstick for estimating the amount of energy consumed. Only heat does. So, when it comes to estimating energy consumption, HEAT (and not motion or light) is our guide.

Stirring in a pan

The table on the previous page illustrates the efficiency achieved in various conversion processes. The efficiency R is defined as the ratio between the energy in the desired form and the total energy used in the conversion process. For power plants, we assume that the heat released to the cooling system remains unused. For electrical appliances, for example a hot plate, the power-plant efficiency has already been taken into account and is taken at 35%. This is a reasonable average for power plants fuelled by coal, natural gas, oil or uranium. This does not apply, of course, for directly-generated electricity, such as with wind turbines or solar cells. For the conversion of solar energy into biomass the year-round average efficiency has been used. In other words, it refers to the harvest taken over a full year. The efficiency of the underlying photosynthesis process is a function of many variables, as we will see in chapter 8.

Conclusions regarding Energy as a Concept

- Energy comes in various forms, including work and heat.
  Its standard unit is the joule (J).
- The various forms of energy can be converted into one another. The important law here is the Law of Conservation of Energy. This law states that the total amount of energy is constant.
- The end product of a series of conversion processes is almost always heat.
- Strictly speaking, energy is never *consumed*. It is only *converted*.
- In the international system of units, energy conversion per unit of time is expressed in watts (W). 1 W = 1 J/s.
- A kilowatt-hour (kWh) is a practical unit of energy. It is energy being converted at a rate of 1 kW (or 1,000 W) for a period of 1 hour (or 3,600 seconds): so, 1 kWh = 3.6 MJ.

3 | Human Energy

Technically speaking, we could see ourselves as engines. The only problem is: we can't turn the engine off. Even when we are not doing any work, we use energy. We are 'idling' so to speak, and need a bit of energy to keep our system going. For example, our heart takes about 5 W to pump blood through our bodies.

When our body does engage in physical labour, the efficiency is limited: part of the energy is converted into heat. It is useful to consider the two – work and heat – separately.

A | Heat

How much do we actually produce, as a heater?
This is easy to trace via our daily food intake. As an adult, we consume about 10,000 kilojoules per day (or, in old-fashioned units 2 to 3,000 kilocalories per day) if we do not engage in much physical labour. If we assume that our metabolism is working properly and our weight remains constant, all that food is converted into heat. This happens during all 24 hours of a day, *i.e.*, 24 × 3,600 seconds. This means that the heat production *per second* comes to just above 100 J/s, that is 100 W. If we exclude top athletes or others engaged in heavy physical labour, our conclusion is:

A human being is a 100-watt heater

as long as the person doesn't engage in heavy exercise. For example, if there are 1,000 people in a concert hall or theatre, they are the equivalent of a 100 kW heater turned on. Needless to say, the air-conditioning system must take this large heat load into account.

When we engage in physical labour, the above-mentioned 100 W of heat increases, of course. This will be addressed in greater detail below.

It is fun to look at the adult energy consumption of 10,000 kJ per day in a different way. We could also express it in oil or petrol. These have a combustion energy of 40,000 kJ per kg. So, in terms of energy, a quarter of that is enough to live on for a whole day. We can conclude that if oil were tasty and digestible we could live on one glass of oil per day. On the highway a car travelling at 120 km/h consumes this in about two minutes. If you look at it from this angle, a human body is pretty efficient.

You can also look at it from a different perspective. Of all the food components we eat (carbohydrates, proteins and fats), fat is the one with the highest combustion energy. And since fat is, roughly speaking, a kind of thick oil, its combustion value is practically the same as that of oil. This means that we could also live on the energy of a 1/4 kg of fat per day. Vice versa, if we stop eating altogether, our energy would derive from the burning of 1/4 kg of fat per day. Slimming products that have us believing we can lose more weight than that should not be trusted – certainly not if we go on eating as usual.

B | Work

Let's take a guess at how many watts of mechanical power a human being can produce. Obviously, the 'human engine' is nothing like that of a car. Just try to push a car and you will quickly see the difference. So, as far as power goes we have much less of it than a car engine, which is typically 50 kW for a medium-sized car.

But how much would this be: 10 W, 100 W, or even 1 kW (the power of a vacuum cleaner)?

This is surprisingly easy to determine if we look at the simple yet efficient exercise of climbing some stairs or riding a bicycle. In these two cases we use the same muscles, and the motion of our legs is almost the same. But walking up stairs is a much easier case to calculate, since it is essentially increasing our potential energy in the gravitational field. All we need to know is the weight of our body, the step height and the step rate. One step per second seems a reasonable rate when we climb stairs for a long time. If this seems a bit slow, compare it to hiking in the mountains: one step per second corresponds to climbing 500 metres vertically in one hour, and that is pretty heavy exercise. So, one step per second is a reasonable starting point.
If we do the calculations (see box) we come up with about 100 W. So we may conclude:

A human being is a 100-watt engine if working on a continuous basis.

This happens to be the same value that we found above for the heat production of a person who does not do any physical labour. But now we are talking about purely mechanical energy. The *total* energy use is a lot higher because the efficiency of our muscles is nowhere near 100%. Much more than those 100 W is released as heat (see *How do I keep my temperature constant?* next heading).

How much – or how little – 100 W is becomes clear if we want to generate electricity using an exercise bicycle equipped with a generator. We do not have to worry about the conversion to electricity: the conversion of mechanical to electrical energy can be highly efficient, *e.g.*, 90% (see table p. 17). For convenience, let us put the efficiency at 100%. Now, if we spend a 10-hour work day generating electricity on an exercise bicycle, at the end of the day we will have produced 100 W × 10 h = 1,000 Wh or 1 kWh of energy. We should realise that we can get that from the wall socket for less than € 0.20.

We conclude that even though electricity seems expensive, pedalling on an exercise bicycle is not the way to go either.

Climbing the stairs

The potential energy we gain by raising our body over a vertical distance h in the gravitational field equals $m\,g\,h$, where m is our mass and g ($\approx 10\ \mathrm{m/s^2}$) the acceleration due to gravity ($m\,g$ is therefore our weight in newtons). Assuming a step height of 15 cm and a mass of 70 kg, the work to take one step is:

$$m\,g\,h = 70\ \mathrm{kg} \times 10\ \mathrm{m/s^2} \times 0.15\ \mathrm{m} = 105\ \mathrm{J}.$$

If we do this at a rate of one step per second, the power is

$$P = 105\ \mathrm{J/s},\ \text{or}\ 105\ \mathrm{W}.$$

So, the mechanical power of a person doing work is typically 100 watts.

Climbing the stairs

C | How Do I Keep my Temperature Constant?

If we are engaged in physical work such as rushing to catch a train or cycling through town, we produce not just mechanical energy, we also produce a lot of heat. Our muscles are not capable of converting food into work with an efficiency of 100%. In that sense, we behave more like a car because most of the energy is released as heat.

Exactly how efficient our muscles are depends on the kind of work we are doing. Weight lifting, for example, is much less efficient than climbing stairs or cycling. These latter two activities have an efficiency of about 25%. This means that the other 75% (or three times as much) is released as heat.
How do we keep our temperature so nice and constant? Of course, we could do this by changing our clothes all the time. Not only would that be impractical, but it would also work only in cases where the ambient temperature is below our body temperature of 37°C. To keep our body temperature constant, we must constantly match the heat *produced* with the heat *released*. As soon as we start working and thus producing more heat, we must increase the release of heat.

There are three possible mechanisms for releasing heat to cool the human engine: heat conduction to the surrounding air, radiation into the environment, and evaporation.

Radiation may sound a bit strange: we never see anybody shine. Nevertheless, radiation is about as important a mechanism as conduction is, but we cannot see it because it is a part of the electromagnetic spectrum that is invisible to our eyes. The wavelength involved is around 10 micrometres (or 0.01 mm), whereas our eyes can only detect wavelengths between 0.4 and 0.8 micrometres.

Losing weight by exercising

Playing sport is a good way to lose weight, in addition it is fun. What is the exact effect of exercise on losing weight? An easy way to look at it is to consider mountain climbing, or cycling, because the two are pretty comparable if we look at muscle activity.

It suffices to look at a simple case of climbing stairs. Climbing one floor requires about 2.5 kJ of mechanical energy (see the box 'Climbing the stairs'). Since muscle efficiency is 25%, about four times as much energy is needed in total, or 10 kJ. This corresponds to about 0.1% of our daily diet. So, to climb a 1,000 metre high mountain – roughly the equivalent of 300 floors – we need about 3,000 kJ, or 30% of our daily food, or approximately one extra meal.

What about losing weight, then? We could skip a meal, or we could do 3,000 kJ worth of exercise, as described above. From the combustion energy of fat, we know that 3,000 kJ is equivalent to burning 75 grams of fat. Suppose we exercise for two weeks in a row, we would lose about 1 kg. And if we feel we have lost more weight than that, it's probably because we haven't been drinking enough!

Temperature profile near our skin

In an ordinary room at home the temperature is fairly homogeneous, even when it is cold outside. The reason is that the air can – and will – flow freely to balance any temperature differences. The air is set into motion automatically as soon as there are temperature differences: warm air will rise, move along the ceiling toward the other end of the room and replace the cooler air. This is called 'free convection'. This mechanism makes the temperature in a room reasonably homogeneous. But, if we look very closely, the convection cannot take the air into the walls or the windows, nor can it flow into our skin. It can only flow parallel to any solid body that is impermeable to air. That is why there is a temperature jump close to our skin. This is illustrated in the figure below.

Why is there such a leap in temperature? It has to do with the poor thermal conduction of the air compared to our skin (or glass, or concrete, for that matter). It is much more difficult to maintain a temperature difference in our skin than in the air. More precisely, we can apply the equivalent of Ohm's law ($\Delta V = I R$, with ΔV the voltage, I the current and R the resistance) which, for this case, is $\Delta T = I R_{th}$, or temperature difference equals the heat flow times the thermal resistance. So, if the heat flow I is the same within the two media, the temperature difference across the layer with the largest thermal resistance (air) is larger than an equally thick layer with a smaller resistance (skin).

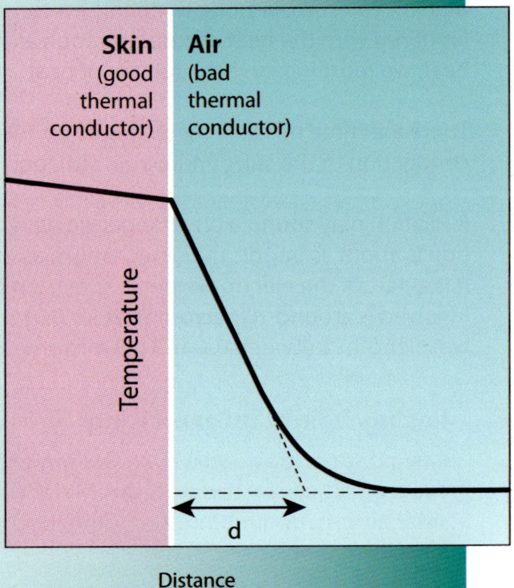

Temperature profile at the interface of our skin (left, good thermal conductor) and the air (right, poor conductor). Beyond a distance d, convection homogenises the temperature.

The steep temperature profile in air holds for the region where the air flows parallel to our skin (d in the figure). This region is 3 mm thick in still air. This value becomes less when the wind blows. Accordingly, the temperature profile becomes steeper, and the heat flow by conduction becomes larger, so we get colder. We can, of course, increase that layer artificially by simply putting on a sweater.

These kinds of temperature profiles show up fairly frequently, for example, near windows (see figures on p. 34).

What about the wind? Doesn't wind contribute to the cooling process? We may be tempted to consider this as a separate mechanism. However, the only thing wind does is transport the cool air closer to our bodies. It can't flow *into* our bodies, because our skin is not permeable to air. All wind can do is flow parallel to our skin. In terms of heat transport, this is the equivalent of a layer of still air close to our bodies (see box 'Temperature profile near our skin'). So, ultimately, conduction has to take over in the last few mm. When the wind blows, it brings the cool air a bit closer to the skin. As a result, the temperature profile near our skin (the 'temperature gradient') becomes steeper, and consequently the heat transport by conduction becomes larger.

The conclusion is that, if we look close to our skin – or to any solid body – we can forget about convection, and we are left with conduction and radiation as the only true heat transfer mechanisms.

But wait: How can these be capable of keeping our temperature constant? Both conduction and radiation are completely determined by the temperature difference between our body and the outside world. They are governed by laws of physics, so there is no way we can influence them.

Fortunately, there is also evaporation. Our bodies are able to influence this process by regulating the amount of sweat we produce, and it does so automatically. To see how this works, let us look at the simplified case of when we start doing exercise without changing our clothing. Think of an exercise bicycle, for example.

The graph shows what happens for various energy and heat components. Muscle power is on the horizontal axis. The total energy production per second and the heat produced per second are on the vertical axis (on an adapted scale). If no work is being performed, both are 100 W. As soon as we start doing work, the *total* energy production goes up about 4 times as fast as the work done. This is because the muscle efficiency is assumed to be 25%. The heat produced is the total energy produced minus the work done.

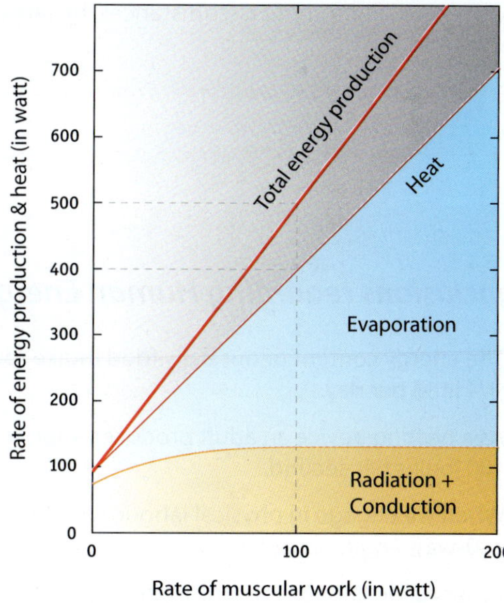

Rate of energy production, heat production and heat released for an adult at work, schematically. Since radiation and conduction are fairly constant, evaporation turns out to be our saving grace.

The *removal* of the heat is broken down into two components. The yellow area represents the heat released by conduction and radiation.

The transport of heat via conduction and radiation remains almost constant, as long as our body temperature remains constant at 37°C. It increases only slightly as we begin working, since the temperature of our skin rises slightly because of enhanced blood flow. However, there is a limit to this: it cannot exceed the core temperature of our bodies, which is 37°C. So, evaporation serves as our saving grace. It fills the gap and plays the key role in the (quite subtle) thermoregulation mechanism, which keeps temperatures constant. Moreover, evaporation is an extremely effective mechanism, because evaporating water requires a lot of heat. Just think of the whistling kettle in the kitchen. If it takes, say, 5 minutes for water to reach the boiling point, it may take another 20 to 30 minutes to completely evaporate. However, during this time, the heat supply remains roughly constant. We can conclude that it takes about 5 times as much heat to evaporate water than to raise its temperature by 100°C!

For dogs, thermoregulation is always a problem. Their fur makes evaporation difficult, so they have to rely on enhanced breathing to eliminate the heat. Dogs pant to cool themselves off.

This is a lesson for people who swim in warm water. Evaporation does not work in this situation. When the water temperature reaches 37°C, conduction and radiation no longer work either. Only the tiny portion of the body sticking out of the water is capable of cooling off. So it may not be such a good idea to try and break a long-distance world record under these circumstances, because the body would get overheated pretty soon.

Conclusions regarding Human Energy

- The energy content of our daily food intake roughly equals that of a glass of oil (1/4 litre) per day.
- As a heating device an adult produces approximately 100 watts, or 100 joules per second.
- When we engage in physical labour, we produce the equivalent of a 100-watt engine on a long-term basis.
- Evaporation is THE mechanism to keep body temperature constant.
- The process of evaporating water requires a surprisingly large amount of energy. Evaporation takes roughly 5 times as much heat as raising the temperature of the same amount of water to the boiling point.

4 | Domestic Energy

A | Domestic Energy Supply

Domestic energy use varies tremendously across the globe. This applies not only to the *amount* of energy, but also to the form in which energy is supplied to homes. In rural areas of developing countries, for example, wood is often the most important (or even the only) source of energy. Cooking is mostly done on wood fires, while home heating is often unnecessary, given the climate. Electricity consumption is low, since it is often used for lighting purposes only.

In contrast, in industrialised countries electricity is the most vital form of domestic energy, since it powers all domestic appliances. Electricity consumption is therefore high. In addition to electricity, a lot of energy is used for private transport (see next chapter) and for keeping the home temperature within a comfortable range. Also here we see large differences across the globe. Whereas most countries in Western Europe and North America spend a great deal of energy heating their homes, many developing countries might be more concerned about cooling and air-conditioning. We will return to air-conditioning on page 29.

To have a better feeling for domestic energy use in an industrialised country, we will look at the situation in the Netherlands.

Energy in most households is used in the forms of natural gas and electricity. How much of each is shown in the following graph, for the period 1980 to 2006. When interpreting these data we should bear in mind that the average number of people per household has declined somewhat during this period.

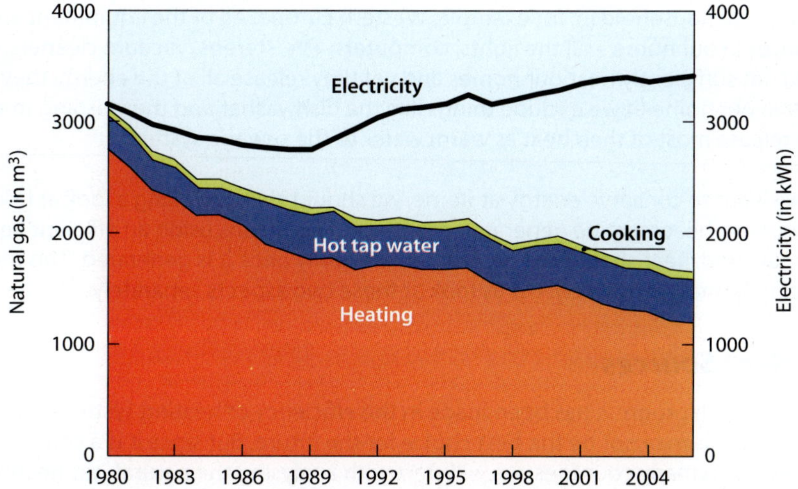

Consumption of natural gas (coloured fields and left-hand scale) and electricity (black line, right-hand scale) for an average Dutch household per year, 1980–2006 (www.energie.nl)

Before we look at natural gas and electricity consumption separately, let us first note two interesting aspects.

- In 1980, the number of m^3 of natural gas consumed by the average household was – quite accidentally – equal to the number of kWh (kilowatt-hours) of electricity: about 3,000. Since then, electricity consumption has risen to 3,500 kWh per year (or 10 kWh per day). In contrast, the consumption of natural gas dropped by half during this period, to about 1,600 m^3 (or 5 m^3 per day).
- We may now be tempted to conclude that natural gas is no longer the most important energy source. But this is not the case. Despite the considerable decrease in consumption, natural gas remains the leading energy source. To get a better picture of this, we must treat the two energy forms equally. If we convert 1 m^3 of natural gas into electricity, it yields about 4 kWh (we assume that the waste heat produced in the conversion process is indeed wasted, *i.e.*, we assume no 'co-generation'). So, the above 1,600 m^3 of natural gas corresponds to 6,400 kWh, which is roughly double the electricity consumption.

Let us take a closer look at natural-gas consumption. As seen in the graph, it has declined over the period 1980–2006. Here we primarily see the result of home insulation and the use of energy-efficient heating systems, since it is the portion used for heating that has gone down so dramatically. In contrast, the (much smaller) part used for hot water production has gone up slightly, while the (even smaller) part used for cooking has remained roughly the same. Note, however, that hot water still requires a lot of energy: on average about one-quarter of total natural gas consumption. This is not surprising considering what we have learned in relation to the law of energy conservation. An easy way to estimate the energy used by a hot-water tap is shown in the box 'Counting flames'.

B | Home Heating and Cooling

When we think about it, it is obvious that heating is the largest consumer of energy in an average household in, for example, Western Europe. All of the equipment and appliances in our home – all the lights, computers, TVs, stereos, vacuum cleaners, etc. – could not sufficiently heat our homes and yet they release all of the energy they use as heat in our home (if we exclude things like the dishwasher and the washing machine which release most of their heat as warm water to the sewage system).

So, if we want to conserve energy at home, we should start by taking a look at home heating. First, heat must be generated efficiently. This means using an efficient heat source. Second, heat should not be lost unnecessarily once it is generated. This means insulating homes adequately. Let us look at these two aspects separately.

B1 | Heat Sources

Although much progress has been made in the efficiency of heating systems based on natural gas, this may not be the best choice for the future. But since it is a commonly used system in many countries, we will start with analysing the natural-gas heating system.

Counting flames

A candle, a burning match, a tea warmer, and an old-fashioned pilot flame: they all use about 100 watts of energy each. So they all produce 100 watts of heat. This yields a nice rule of thumb:

A single flame represents about 100 watts of power.

Of course, not all flames are equal, and even candles can vary from one to another. Even so, a flame is a good yardstick when we seek insight into the most important energy consumers in our homes.

Take a gas stove, for example. How much power does it represent? Look at how many flames a single gas burner contains: about 20 to 30. So the power of a gas burner must be 2 to 3 kW. If we search on the Internet for an arbitrary brand of gas stove, lo and behold, we find almost exactly those figures: 1.75 kW for the average burner, 2.7 kW for the larger one and 3.5 kW for the so-called 'Wok burner'. That's beautiful! The figures for an electric cooking plate must be roughly similar – and they are – since cooking on an electric stove takes just as long.

Let's take another step: the warm-water tap heated by an electric boiler or a gas geyser. If you look inside a gas geyser, you will notice that it contains about ten rows of ten flames each. This makes 100 flames at 100 W each, so we expect a heating power rating of 10 kW. That may sound like a lot, but if we search the Internet, the first kitchen geyser we find quotes a power rating of 9.4 kW. Long live our rule of thumb!

By the way an electric boiler in the kitchen is unable to continuously supply such a high power: a standard 16-Amp fuse would blow immediately. Just check: current times voltage is power, so 16 A × 230 V = 3.7 kW. This is much too low to supply the 10 kW needed. Therefore, an electric geyser is, in fact, a boiler: it must collect heat over a certain period of time to produce a fixed supply of hot water to compete with the gas geyser. And, even then, the result may still be too poor to produce a decent shower.

For filling a bath tub quickly (or taking a comfortable shower) we obviously need something heavier than the kitchen geyser. We expect it to be 20 to 30 kW. A central heating unit may have even more power, depending on the type of house.

One conclusion is that a hot-water tap uses about 10 kW of energy, if not more. So, it is a huge power user. If the water is heated electrically, we can directly compare this to lighting. Take, for example, traditional incandescent light bulbs with an average of 50 W. Comparing this to the 10 kW above, we find that we can turn on 200 of these light bulbs using the same power as a modest electric water heater. If we turn to more efficient lighting, like fluorescent lights (TL), energy-saving lights or LEDs, we would come up with a number close to a thousand. In short, if I use the hot-water tap during a certain time, I might as well switch on a few hundred to a thousand lights during that period of time and use the same amount of energy!

Another conclusion we can draw is that natural gas is an extraordinarily handy fuel for domestic purposes, especially when it comes to high power. Heating water, for example, may be difficult if we want to use electricity. Furthermore the original conversion of heat to electricity at the power station remains inefficient (see table on p. 17).

Natural-gas heating system: Modern 'high-efficiency' heating units are very efficient indeed. This is achieved, first, by a heat exchanger which recovers heat from exhaust gases. The principle of a heat exchanger is explained on p. 30. Moreover (and most importantly), use is made of the heat gained from the condensation of the water vapour. The combination of these features leads to efficiency ratings of nearly 100%.

Heat pumps: There is a way to achieve efficiency ratings of way over 100%. Strange though it may seem, this does not violate the law of the conservation of energy (of course not!). This is because it only *moves* energy around rather than converting it into heat. The word 'efficiency' is thus misleading. Therefore, the term 'Coefficient of performance' (COP) is often used. It refers to the amount of heat moved divided by the energy used in the process. The trick is that the amount of heat moved from one spot to another is generally much larger that the energy needed to do so. In fact, we may be aware of heat pumps because we have one in our refrigerator. All it does in this case is transport heat from inside (which is cooled by evaporation of some fluid) to some sort of black grid behind the refrigerator (which heats up as a result of condensation).

Now, suppose we remove the refrigerator door and mount the refrigerator *backwards* in a window. This means that the black grid will now heat our room while the cold inside of the refrigerator is exposed to the outside air. The (relatively warm) outside air will make the inside of the refrigerator less cold. In other words, it supplies heat to the unit. This heat, in turn, is 'pumped' into the house and released by the black grid.

The energy needed to drive this cycle is supplied by a small motor, usually an electric one. Since the energy used is much smaller that the heat transported, this is an extremely efficient heating method. Indeed, COP values can easily be 3 to 5 if the process is optimised. However, this depends on the temperature difference between inside and outside, *i.e.*, on the temperature step we have to make. The larger this step, the smaller the efficiency will be. In fact, the theoretical value of the COP is $T_{high}/\Delta T$, the ratio between the higher temperature, in kelvin, and the temperature difference. It is, therefore, desirable to have a minimal temperature differential. In practice, this usually means finding a heat source (at the lower temperature) that is as warm as possible. An option would then be to draw the heat from a large body of water, *e.g.*, from a relatively warm underground reservoir. This is one application of Geothermal Energy (see chapter 10).

We can conclude by saying that the future probably looks bright for heat pumps.

Heat from co-generation: The best system involves heat that would otherwise be lost. This can be done by using the waste heat from power plants, which would normally disappear into the cooling water and end up, for example, in a river. This system is called 'Co-generation'. However, to use this 'cooling water' for heating purposes, its temperature must be at a reasonably high level, on the order of 80°C (if used in standard radiators) or 40°C (if used in floor heating systems). This lowers the electricity generation efficiency, since it affects the first step: heat → mechanical energy (see chapter 2 and the table on Conversion processes, p. 17).
The concept of co-generation is especially useful if the heat can be utilised close to a power plant. This is the case, for example, in small urban power plants.

Micro co-generation: This is a small-scale co-generation concept specifically for homes. Here heat production is the primary goal, and electricity is generated as a by-product. It is, so to speak, a natural-gas fired central heating unit that heats the house and produces some additional electricity in the process. It uses the high temperatures that are reached in the combustion of natural gas to produce electricity. This results in part of the energy being converted into electricity and the rest being used for heating. Since electricity is a much more useful form of energy than heat, this approach can enhance the overall efficiency and also save money: energy in the form of electricity is more expensive than the same amount of energy in the form of heat.

The pay-back period for these systems is usually 5 years. The engine used to drive the generator may be a Stirling engine, which needs little maintenance. There is more information about the Stirling engine in chapter 8, p. 110.

Solar collectors: We will discuss solar collectors and their efficiency in chapter 8. They have a drawback for home-heating systems because most of the heat is collected in the summer when we need it the least. This means that long-term heat storage is necessary, which, in turn, requires large storage units (see chapter 12).

Passive solar heating: This term is often used for solar home heating without employing technological devices. This is a largely underestimated field. We should realise that glass windows transmit solar energy quite efficiently. Of all incoming solar radiation at normal incidence (perpendicular to the surface) only 8% is reflected by single panes and about 15% by double panes. Since absorption is relatively small, the majority of the incoming solar radiation can be used for home heating.

B2 | Cooling and Air-Conditioning

Traditional cooling and air-conditioning units operate on the principle of a refrigerator, *i.e.*, of a heat pump discussed on p. 28. It usually uses some fluid that needs to evaporate on one side of the unit (the cold side) and condense on another (the hot side). This involves some kind of pump that circulates the fluid through the unit.

Air-conditioners are often large consumers of electricity. It is therefore important to know that electricity consumption increases sharply with the temperature difference to be bridged. This is because, first, more heat has to be removed from the home or office when the inside-outside temperature differential increases. Second, the efficiency of any heat pump *de*creases when the temperature differential increases (see p. 28). It is, therefore, wise to maintain as small a temperature differential as possible. This can be done in two ways. First, keep the desired temperature reasonable, and not too low. Second, make sure that the hot end of the unit can get rid of its heat efficiently, by keeping it clean and well-accessible by fresh air. This may also be achieved by cooling it with water.

As with heating, the possibilities of passive cooling methods are grossly underestimated. Blinds and awnings can keep the solar heat out at no energy cost whatsoever. It will become clear in the section on radiation (p. 33) that we must intercept the solar heat *before* it penetrates the windows.

The principle of heat exchangers

Suppose you are taking a shower. While you are in the shower, you think about energy and suddenly realise that the water you are using enters the home at 10°C and is heated there to about 40°C. Meanwhile, the same amount of warm water at almost that same temperature is literally going down the drain. Can't we heat the incoming water with the outgoing water?

Good question. Of course we can. If we are smart, we can even do it at almost 100% efficiency, by using the 'counter-current principle' (see figure). We make use of that principle without realising it, when we use central heating. Warm water enters a radiator from above, and leaves as cold water below. In contrast, the air that needs to be heated enters at the lower (relatively cold) side and exits at the top of the radiator. In this way the rising air is heated gradually, and ultimately picks up the temperature of the hottest part of the radiator: a very smart idea indeed.

The difference between the 'stupid' heat exchanger and the counter-current exchanger is shown in the figure. In relation to the shower, we put the incoming and outgoing pipes next to each other and increase their thermal contact. If the two flows were in the same direction ('concurrent'), the water coming out at both ends would be lukewarm, at the average temperature of incoming and outgoing water, see left-hand panel. But if they were to flow in opposite directions ('counter-current'), we would increase the efficiency. See the panel on the right: the flows will exchange their temperatures almost completely!

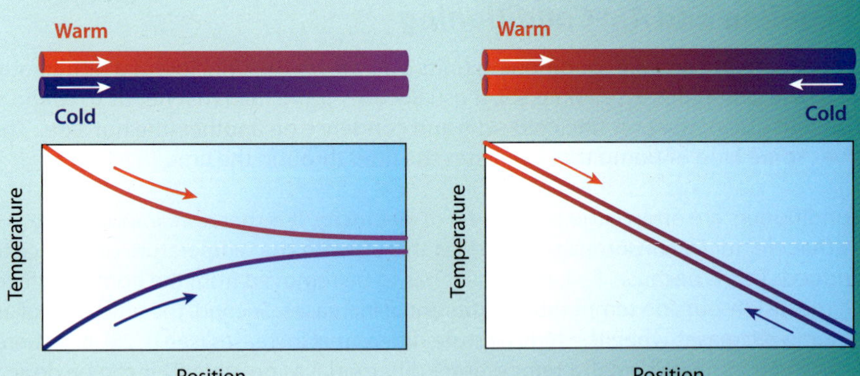

Granted, it may not be that simple, because, in this case, the outgoing water may be dirty and deposit a layer of dirt on the interior of the pipe, thus decreasing the thermal contact. But, in many other cases (and in the case of air), this may not be a problem at all.

Heat exchangers

Moreover, it is wise to use white or, even better, shiny metal surfaces to reflect as much of the incoming heat as possible. This keeps the blind relatively cool which means that, in turn, its emission to the windows is kept to a minimum.

B3 | Heat Exchangers: Free Energy

We can also 'recycle' heat once it has been generated inside the home. The idea is to use heat exchangers. Think of hot water leaving the house. It may be dirty, but it does contain a lot of heat. We can re-use this heat by having it warm up incoming cold water, so it does not need to be heated as much when the hot water tap is turned on. This can be accomplished with almost 100% efficiency if we utilise the counter-current principle. In fact, counter-current is a general principle which is frequently used also in the human body wherever efficient exchange of some kind is required, for example in the kidney.
The principle of heat exchangers is outlined in the box on page 30. Note that a heat exchanger can be used both ways: to recover heat or, if used the other way around, to recover cold.

B4 | Heat Losses and Home Insulation

Once we have generated heat, we should use it efficiently to heat our homes. An interesting detail is that we don't need to know how much energy it requires to heat up our rooms because, normally, the only thing we need to do is *keep* our home warm by compensating for heat loss. This means that we do not necessarily need to know the so-called 'heat capacity'. It does play a role if we are interested in the speed at which a home heats up or cools down. It also determines the power that the heating unit must produce in the morning to heat up our home, assuming that we maintain a lower temperature during the night. But for the overall energy use it is irrelevant.

So we must look at the *losses* of heat, which determine the amount of energy necessary to keep a home warm. Minimising heat losses means increasing insulation.
Since convection, conduction and radiation are the chief mechanisms for heat transport, we will now consider them in greater detail.

Convection, more specifically *natural* or *free* convection, is the motion of air owing to differences in temperature, and thus in density. It serves to homogenise room temperatures, as was already discussed in chapter 3 (see box on p. 22).
The pleasant role that convection plays also has its drawbacks. For example, it causes a draught near cold windows: the (relatively heavy) cold air will flow down and into the room, making our feet colder. This is especially true near single-pane windows if there is no heater in front of it. Another drawback of free convection is that it limits the effective width of the insulating air gap between double-pane windows. Only the thin layers of air next to the glass surfaces contribute to the insulation. Each of these layers is a few mm thick. So, if the air gap is wider than, say, 1 cm, the remainder of the gap does not contribute to the insulation, because convection takes over in that region and tends to make the temperature homogeneous. The onset of convection in such a case is visible in the panel on the right in the figure on p. 34.

Conduction takes place even in a medium that is completely at rest. It is caused by the thermal motion of the individual atoms or molecules, as well as the free electrons in the case of electric conductors like copper or iron. These particles pass the thermal energy along. The *amount* of heat leaking through a window or a wall is proportional to the surface area. Furthermore, it is proportional to the temperature difference between the front and the back, and inversely proportional to the thickness of the wall or window. (Note that the two latter quantities can be jointly expressed as the 'temperature gradient', or the steepness of the temperature profile.)
Ultimately, the thermal properties of the specific material come into play. They are expressed as a single quantity: the *thermal conductivity coefficient*, often written as λ (the Greek letter lambda). So, for any given geometry, the higher the value of λ, the higher the heat loss as a result of conduction.

The value of this coefficient varies tremendously for different materials. As a rule of thumb, one can say that its value is very small for gases, intermediate for liquids and relatively high for solids. Metals are the best conductors. The reason is that the free electrons, which transport the electric current, also transport heat in the process.

To get an idea of the value of the thermal conductivity, its approximate value is shown for a number of selected materials in the table below. (The amount of joules per second (or watts) leaking through a layer of a material equals $A\lambda\ \Delta T/d$ where A is the surface area of the layer, ΔT is the temperature difference and d is the thickness of the layer.)

When we look at this table we notice a few things.
- Air is a very poor conductor compared to glass or steel, for instance. That is, as long as convection does not play a role (see, for example, the box 'Temperature profile near our skin', p. 22). This means that, for all practical purposes, air is basically the best insulator we have (besides a vacuum and a few rare gases). The excellent insulating properties of various types of plastic foam derive, in fact, from the air that they contain. The only role that the foam plays is in keeping the air from moving around and causing heat loss via convection.

Thermal conductivity coefficient λ for various materials at ambient temperature. A low value means low heat loss (or good insulation). The unit is W m^{-1} K^{-1}.

Xenon	0.005
Krypton	0.009
Argon	0.016
Air	**0.025**
Oil	0.1
Water	0.6
Wood	0.4
Glass	1
Steel	50
Brass	120
Aluminium	240
Copper	400

- Glass is a much better conductor than air. At an air-glass interface, we thus find a similar temperature profile as seen in the figure on p. 22. The reason for this is that the air near the interface can only flow parallel to the surface. Therefore, we have a layer of a few mm thick that can be considered as still air. Now, since glass is such a good conductor compared to air, the temperature across the glass is almost homogeneous. The temperature differences only occur within the thin layers of air next to the glass.

From the above we can conclude that we cannot obtain any improvement in insulating values by increasing glass thickness. An extremely thin layer of glass performs almost as well as a thick one, as do curtains that are drawn tightly and leave little space for the air to pass through. After all, as long as we have a layer that is impenetrable to air, the heat will not be transported. It is the air layers that do the job. They form the insulating barriers between warm and cold.

The above also explains the temperature profile near a window.

The ***single-pane*** case is illustrated on the left of the figure on page 34: the temperature jumps occur in the two air layers next to the window.
One may wonder why the temperature jumps are not equally large. This is because of the influence of the wind. On the outside of the window, the wind makes the insulating air layer a bit thinner. Consequently, the glass temperature is 'drawn' toward the outside temperature, so the glass is colder than one might expect. The figure illustrates the situation for an average wind speed.

The ***double-pane*** case is illustrated by the panel on the right. In this case, there are four instead of two layers of air, so that the insulation value is roughly doubled.

Double-pane windows have, in fact, *two* advantages. Not only are the heat losses cut in half (and so is the heating bill), but there is also an advantage regarding comfort. This is the direct consequence of the fact that the inner pane has a higher temperature: +8°C compared to –1.5°C in this example. The draught of cold air near the window, which can be quite uncomfortable, is dramatically reduced, because the air speed is slower *and* the air temperature is higher. Furthermore, heat losses via radiation are lower for double-pane glass, which is especially noticeable when one stands near a window.

The conclusion is that double-pane windows have *two* advantages (apart from noise reduction): less energy loss and greater comfort.

Radiation always plays an important role. In most cases, its contribution is just as important as conduction. We noticed this when we studied the thermal regulation of the human body (p. 21). This can easily be verified by calculating the heat conduction through a 3-mm layer of air, and comparing the result with radiation losses for the same temperature difference. In the latter case, Stefan-Boltzmann's radiation law is used (see chapter 8, p. 92), keeping in mind that we must measure the difference between the incoming and outgoing radiation.

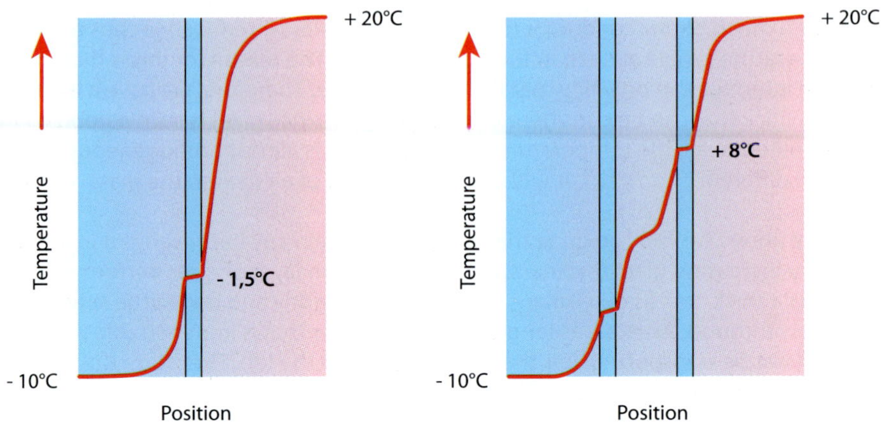

Typical temperature profile near single- and double-pane windows, given an outside temperature of –10°C and an inside temperature of +20°C. The slight kink in the centre of the double-pane case marks the onset of turbulence, which tends to make the temperature profile flat.

An interesting feature of radiation losses is that *each* individual substance emits radiation, also at room temperature. But we do not notice this because our eyes are not sensitive to this part of the electromagnetic spectrum: infrared (IR) radiation with wavelengths around 0.01 mm. This is 20 times longer than visible light. In turn, this is because ambient temperatures (about 300 K) are lower by a factor of 20 than the outer layers of the sun (almost 6,000 K).

This concept is illustrated in the figure on page 35. Because of the large differences between the two wavelength regions, the horizontal scale is a logarithmic scale. The peak of the solar emission spectrum coincides nicely with the peak of human eye sensitivity, which is at 0.55 µm. Note that there is hardly any overlap between the solar emission spectrum and the spectrum emitted by any surface at ambient temperature.

This significant difference in wavelengths has several consequences. Not only is the IR invisible to the human eye, it also has completely different optical properties. Something that looks perfectly white to our eyes (*i.e.*, it reflects the visible part of the spectrum) may be 'black' in the far IR region (*i.e.*, absorbs the IR radiation). Indeed, almost all surfaces are 'black' in that region. They have *absorption coefficients* near unity, even snow. The only exceptions are metal surfaces. These can have absorption coefficients close to zero. For example, aluminium and copper surfaces have absorption coefficients below 0.1.

The *Emission* of a surface is efficient if absorption is efficient. This follows from a balance equation: what comes in must go out in the stationary state. Since most surfaces are 'black' in the far IR region, they are not only good absorbers but also good emitters. For all practical purposes, we can consider all surfaces 'black' in the far IR region around 10 µm. Again, metal surfaces are the exception: they are both poor emitters and absorbers. It therefore makes sense to use a shiny metal teapot, for example, because it loses little heat through radiation.

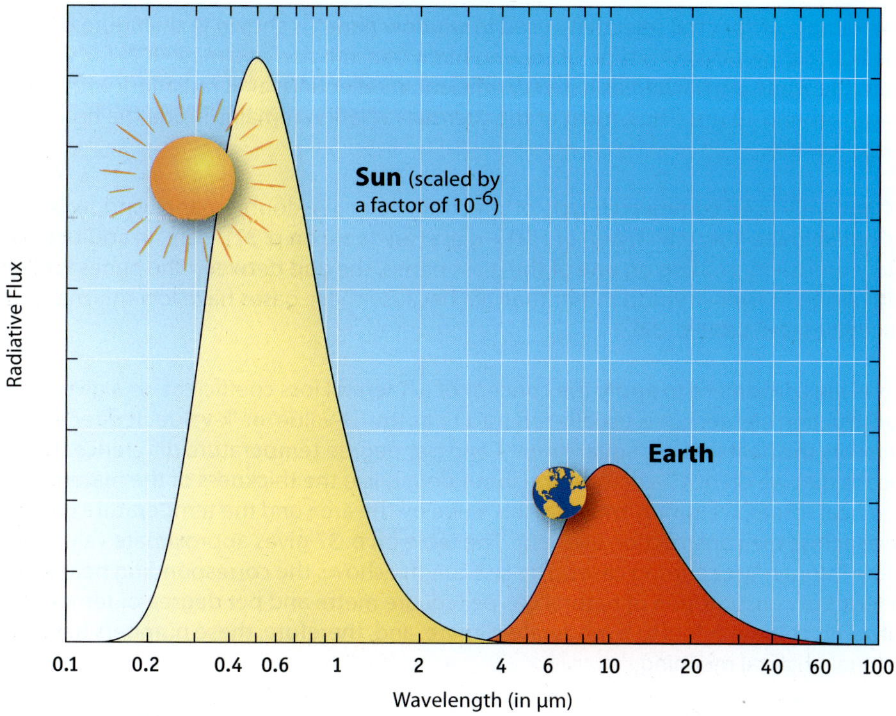

The emission spectrum of the sun peaks at 0.5 µm, at the peak of the sensitivity curve of the human eye (see figure p. 43). In contrast, the thermal radiation of objects with ambient temperatures (300 K) peaks at 10 µm.

Similarly, it makes sense to use aluminium foil as a thermal shield behind a radiator positioned against a cold window or a cold wall. It reflects 95% of the radiation and absorbs only 5%. White paint, on the other hand is ineffective in this respect, since it behaves effectively as a black surface for this type of radiation. Since this is the case for *all* colours of paint, it also follows that a radiator's colour is irrelevant. It will behave as an almost 100% efficient emitter irrespective of its colour, unless the paint contains metal (*e.g.*, aluminium paint).

We can now take advantage of the difference in optical properties between visible light that is emitted by the sun, and the far end of IR radiation emitted by surfaces at an ambient temperature. This is done in window panes with a spectrally selective coating. These can be manufactured to have a *high* transmission of incoming solar radiation, and a *low* transmission of 'far IR' radiation, which would otherwise escape from a room. This is achieved simply by reflecting the IR radiation. The precise optical properties may vary to suit a specific situation and climate. For example, in a cold climate one may want to have as much solar heat coming in through the windows as possible. Thus, the 'near IR' part of the incoming solar spectrum may be welcomed along with the visible part of the spectrum. This may be different for a mild climate.

An example of spectral selectivity used in window panes is shown in the figure below. The figure shows that this type of coating has a transmission of almost 90% for visible light (the white part) whereas it is very efficient in reflecting far IR radiation, with wavelength around 10 μm. Thus, most of the thermal radiation remains inside the home and keeps it warm.

A frequently used commercial type of spectrally selective double-pane window is the so-called low-e (low emissivity) or HR++ glass (see table on p. 37). Here, in addition to a low-emissivity coating on one of the glass panes, the gap between the panes is filled with argon or even krypton rather than air, because these gases have low thermal conductivities (see table p. 32).

It is often convenient to apply the concept of a 'thermal loss coefficient' to a given construction element. It is usually referred to as the 'U-value' or 'k-value'. It directly presents the heat loss per square metre and per degree temperature difference. This means that we do not have to worry about details like the thickness of the material. To find heat losses, we simply multiply this value by the area *and* the temperature difference across the construction element. The table on p. 37 gives approximate values for some frequently occurring cases. The last column shows the corresponding heat losses in terms of consumption of natural gas per square metre and per degree of temperature. This column refers to a moderate climate, and, therefore, these numbers have no general physical meaning.

It is sometimes useful to use the concept of thermal resistance, which is the inverse of the U-value quoted above. Just like the resistance in Ohm's law for electrical currents, the thermal resistances simply add up for construction elements placed one after the other.

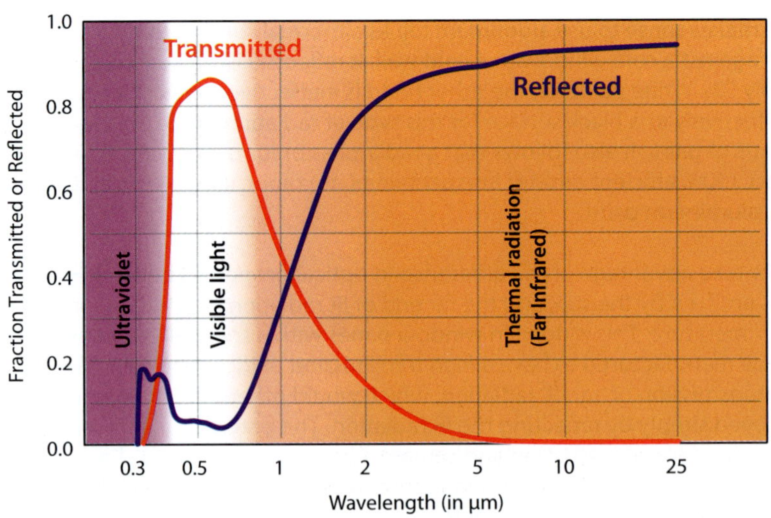

Example of spectral selectivity. This window transmits most of the incoming sunlight while keeping 90% of the thermal radiation inside.

Heat loss in watts per m^2 of a window or wall, per degree temperature difference: the 'U-value' or 'k-value'. The insulation mentioned here serves only to prevent convection, and thus mimics a layer of still air. The column on the right shows the approximate heat loss expressed in m^3 of natural gas per m^2 surface and per year in a moderate climate (*e.g.*, London or Amsterdam).

	U value in W/m^2 K	Heat loss in m^3 of natural gas/m^2 year
Single-pane window	6	60
Double-pane window	3	30
Low-e or HR^{++} window	1.2	12
Hollow wall, 5 cm air gap	2	20
Hollow wall, 5 cm insulation	0.6	6
Hollow wall, 10 cm insulation	0.35	3.5

C | Hot Water

As we saw in chapter 2, heating something requires more energy than most people think. This is certainly true for water. Heating a certain volume of water takes roughly twice as much heat as heating any other liquid or solid like a metal.
Now, if we want to heat up some water, what would be more efficient: using electricity, natural gas or solar power?

Electricity is practical. However, an instant water heater located under the sink is expensive and not very efficient. Granted, almost all of the electricity will go to heating the water, since the efficiency of heat transfer inside the boiler is almost 100%. However, most of our electricity is made in thermal power plants that run on coal, oil, natural gas or uranium. In all of these cases, less than half of the heat is converted into electricity (see chapter 2), and most of the energy is produced as waste heat. So, unless that waste heat is not really wasted (but used for other heating purposes) this is not the best option. A better alternative would be to use *all* of the heat released by combustion, which we do if we burn the gas at home rather than at the power plant.

A *gas boiler/geyser* would be a better – and cheaper – option. The heat transfer within such an open system may be somewhat lower than in an electric boiler, but that does not negate its sizeable advantage. A water-heating system is often part of the central heating unit, which does not jeopardise the high efficiency of the combined unit.

Solar collectors are good hot-water suppliers, because hot water is necessary all year round, so we can easily use the collectors during the summer, when the sun is delivering most of its energy. This is in contrast to home heating, where we need a long-term storage system to achieve any savings. In most cases, a 100-litre storage tank for hot water is sufficient, combined with a solar-collector surface of 3 to 5 m^2.

Anything less than that would not meet the average hot-water demand, and any more would yield more hot water than we need, especially in the summer. So, finding the optimum is always a compromise and depends on the particular details and the geography.

In a moderate climate, such as most of Western Europe, this is not an adequate solution. During cloudy periods a back-up heating source, fuelled for instance by natural gas, is a necessary addition.

A solar hot-water system thus has at least two components: a solar collector and a hot-water tank. In a moderate climate, a third element should be added in the form of an external heater for less sunny periods. We will return to solar energy in chapter 8.

As mentioned earlier, one often underestimates the amount of energy needed to heat water. In the box 'Counting flames' (p. 27) we saw how a running hot-water tap uses 10 kW of energy or more. Suppose we do this for 1% of a day (or a quarter of an hour), we consume an *average* of about 100 W or more. If this is heated electrically, we can directly compare it to the total electricity consumption of an average household. These 100 W would be about a quarter of the total electricity consumption for the average European household. This means that washing dishes under a running hot-water tap is not something we should do too often. Before long the hot-water tap will end up as the leading electricity consumer in the home!

Is taking a bath less efficient than taking a shower? Probably, but this obviously depends on the amount of hot water used. For a specific case, we can compare the time it takes for a tap to fill up a bathtub with the time we spend taking a shower, keeping in mind that showers usually take less water per second than a bathtub filling up. But if we live in a cold climate, the energy from the bath's water can be recuperated by letting the water cool down before draining. This way the heat from the water can help in heating the home.

D | Baking and Cooking

Here we face the same question: electricity or natural gas? In the hot-water case, we have already noted that the answer is natural gas. The answer may be less obvious for baking and cooking because it involves the efficiency of a gas burner heating a pan or a kettle. This efficiency is definitely lower than in the case of heating water. But if we make sure that the size of the gas-burner's flame does not exceed the size of the pan, we can easily achieve an efficiency of more than 50%. In contrast, the electricity in a home represents only about 35% of the primary energy used to generate it in a thermal power plant. So, even if the heat transfer from heat source to pan is efficient (say, 70%), the overall process is less efficient than heating by natural gas. This is also true for other electric cooking methods such as induction heating; their efficiency is always limited by that first step, the generation of electricity.

The microwave oven is a special case. Although it has the same drawback as other electric-heating cases, its advantage is that it is capable of heating only the things we are interested in: the food. So, for small amounts of food that do not require a large pan, the microwave may be the best choice. In addition to being convenient, it is relatively energy efficient.

Solar cooking is a 'green' option especially in sunny developing countries. However, bear in mind that the sun's radiation corresponds to 1 kW per m^2, whereas we usually need a multiple of 1 kW to do any cooking. So we need a collector surface on the order of 1 m^2 or more, concentrate the solar radiation onto the surface to be heated, and make sure we minimise the heat loss. The latter can be easily achieved by using transparent foil, which then acts like the glass in a greenhouse. As we have seen earlier, even a thin foil provides just as much insulation as glass, and, at the same time, it minimises heat loss via evaporation.

A parabolic reflector is an option, because it focuses the parallel sunrays to a single point. However, since the surface to be heated is relatively large, the quality of the reflector curvature is not that critical. Therefore, one can go a step further and use a simple arrangement of curved surfaces covered in aluminium foil for maximised reflection. This may be just as efficient and much easier to make. An example is shown in the picture below.

Solar cooking using a simple reflector and a black pan in a plastic bag (Source: Stichting Solar Cooking Nederland, www.solarcooking.nl).

E | Electrical Appliances at Home

Let's take an average Dutch household as an example. As we have seen in the figure on page 25, annual electricity consumption is 3,500 kWh. Since a year is comprised of $365 \times 24 = 8{,}760$ hours, this corresponds to 0.4 kW or 400 watts continuously. So, the average electricity consumption corresponds to having four 100 W lights on continuously. In an average household using incandescent lighting, this may be roughly what lights use if they are switched on. But wait, they are switched on only 20% of the time, so we can conclude that lighting is not the major source of electricity consumption. There are, of course, many other electrical appliances such as the washing machine, dishwasher, dryer, vacuum cleaner, TV, refrigerator, freezer, stereo, and so on. As it turns out, lighting is only about 20% of domestic electricity consumption if we use incandescent lamps, and much less than that if we use efficient lighting such as fluorescent 'saving lamps' or Light-Emitting Diodes (LEDs). See p. 45 for a comparison between various light sources.

E1 | The Major Culprits

First let us look at the major consumers of electricity in our homes. The law of conservation of energy leads the way. Since energy is almost always converted into heat, we must find those appliances that produce a lot of heat. Or, a category often overlooked, appliances that are 'on' continuously, such as the refrigerator. So the major electricity consumers in an average home are the following, with the approximate annual consumption in kWh:

- Electrically powered hot-water taps, up to 2,000 kWh;
- Washer and dishwasher, 500 kWh each (because they produce hot water);
- Dryer, 500 kWh (produces hot air!);
- Refrigerator/freezer, 500 kWh (is *on* continuously!);
- Lighting, 100 to 500 kWh, depending on type.

Notice that the dryer consumes a lot of energy. This is because it must supply the heat of evaporation, which is surprisingly large (see p. 24). In contrast, removing most of the water from clothes by spinning them costs almost nothing. So it pays to use a high-RPM (revolutions per minute) setting for the spin cycle on the washing machine. An excellent alternative to using a dryer is drying clothes on a clothesline in the fresh air, letting the sun and wind do the job for free.

There is an interesting aspect to electricity use at home. Guided by the law of conservation of energy we realise that, at the end of the day, *all* electricity is converted into heat. This means that all appliances contribute to heating our home (except, of course, for those that simply throw away the heat as waste such as the washing machine and dishwasher's hot water). This is good news for those who live in a moderate climate (but bad news if the air-conditioning must be on).

E2 | Lighting

Although lighting, as we have seen, forms only a small portion of actual domestic electricity use, it is easy to improve lighting efficiency by at least a factor of 5 by switching from incandescent lamps to fluorescent tubes (TL) and energy-efficient light bulbs (both based on gas discharge), or Light-Emitting Diodes (LEDs). The efficiency of these lights is exceptionally high, certainly if we compare it to a candle's efficiency.

Candles

Candles are cosy, no question about it. They add an air of intimacy and romance to a relaxed dinner. They were originally used as smart light sources, much handier than oil lamps while using zero electricity. Some people even insist that it is a super energy-efficient light source.

Can this be proven? How much energy does a candle actually use when it burns? The answer is: much more than people think. We can get an idea by looking at the heat it produces, and that is considerable. An educated guess would be that it produces about as much heat as an incandescent light bulb of 100 watts. That sounds like a lot!

But we don't have to guess: we can easily *measure* the energy use by how much candle wax it uses in an hour. If we combine this with the heat of combustion, we discover that the heat generated is roughly 100 joules per second (see box 'Candle heat'). That is indeed 100 W, while the light produced by a candle is nowhere near that of an incandescent 100-W light bulb, let alone that of a cluster of efficient light bulbs whose power adds up to 100 W.

The number of 'watts' is a measure of the energy used, not the amount of light produced. Take an electric heater, for example. It may consume 1,000 W but nobody would call it a good light source. It radiates primarily in the infrared part of the spectrum, which our eyes cannot even see. If we are discussing light, we mean radiation that can be detected by our eyes, by the photoreceptors – the rods and cones – on the retina. That type of visible radiation is not measured in watts, but in 'lumens'.

Thus, the efficiency of a light source is determined by the amount of light it produces per unit of energy it uses. The efficiency can therefore be expressed in the number of lumens per watt.

When we look at a candle from that perspective, we see that its performance is relatively poor. It produces 0.15 lumen per watt (lm/W). A traditional incandescent light produces 12 lm/W, which is 80 times more. An energy-efficient light bulb and a Light-Emitting Diode do even better. At 60 lm/W, they are 400 times more efficient than a candle!

There is a catch, however. The electricity is usually generated from heat in a power plant, which itself has a rather poor efficiency rating. The electricity we get from the outlets in our home represents only about one third of the total energy that the utility company needs to produce it. But even if we take that into account, electric lights are still far more efficient than candles.

The conclusion is that candles may produce a romantic atmosphere in our homes; they may even help to heat up our homes a bit. But if we need a lot of light, we should simply take them down to the power plant and throw them in the oven, rush home and use the electricity they produce to light an energy-efficient light bulb.

Candle heat

If we want to find out how much candle wax is consumed by a candle, all we have to do is weigh it before and after burning it for an hour or so. Of course, not all candles are equal, and not all flames are the same. But an average candle loses 8 grams per hour.
The heat of combustion for candle wax is 48 kJ per gram. Consequently, those 8 grams represent about 380 kJ of heat, produced in 1 hour. That is 380,000 J / 3,600 s or roughly 100 W.
This is approximately as much as any human being not engaged in heavy exercise expends. We can easily check this by observing that 100 W during 24 × 3,600 seconds, or one day, yields 8,640 kJ. This is a perfectly normal diet for an adult.

How can we *feel* the actual consumption of electricity?

Of course you can measure energy consumption. Or you can read about it on the labels accompanying your electrical appliances. Or you can look up the power rating in the manual. But to get a reasonable estimate none of these methods are necessary. Most of the time we can feel the energy use by the heat that an appliance produces. After all, according to the law of conservation of energy, the energy has to go somewhere and nearly always it ends up in the form of heat. So, a 1,000 W vacuum cleaner is a 1,000 W heater, a 60 W lamp is a 60 W heater, and so on. If we want to know if a DVD player, a mixer or an electric shaver uses a lot of energy we only need to put our hand on them to know. Appliances that remain cool are not big energy users (if they are not running at full capacity 24 hours per day, because even modest energy use will eventually add up over time). If appliances need a cooling fan, they are probably significant energy users. Appliances and systems that are designed to produce heat – the home heating unit, the boiler, the washer, the dishwasher – are the truly big energy consumers.

A nice illustration of estimating energy use is the electric shaver. It just can't use that much energy. Where would the energy go? Heat is the only option: electrical (or 'Ohmic') heat produced in the electrical wires and the motor, and heat generated by friction in the shaving head. But that can't be all that much, because the shaver remains relatively cool during the entire shaving operation of a few minutes. The amount of heat involved is clearly much less than the amount that would be necessary to heat up enough water for an old-fashioned wet shave.

When we do the calculation, we discover that shaving with the electric shaver uses as much energy as that involved in opening the hot water tap for *one* second. So, a wet shave does not save energy: it uses far more than an electric shaver (unless one uses cold water, of course).

Why is an incandescent light bulb so inefficient?

The leap from candle to incandescent lamp was quite dramatic at the time. No more fumbling around with matches, no hot wax spilled on the furniture, no danger of fire, and a lot more light. That said, this old-fashioned incandescent bulb is just not *that* efficient. It uses a lot of electricity and, in return, does not provide all that much light. Why is that?

The light bulb's efficiency is determined by the degree to which the radiation emitted can be sensed by our eyes. So, we must compare the emission spectrum of the bulb with the eye's sensitivity.

This has been done in the accompanying figure. The curve on the left shows that our eye is sensitive only to wavelengths between 400 and 700 nanometres, the colours of the rainbow (one nanometre is one millionth of a millimetre). The peak lies at about 500 nm, in the yellow/green part of the spectrum. That is where our eyes are the most sensitive. On the right we see the spectrum emitted by the incandescent wire in a bulb. Its peak is at about 1,000 nm.

The two curves simply don't match up. The largest part of the light's emission falls outside of the range of our eyes and remains invisible to us. An incandescent bulb with this type of emission spectrum is simply inefficient by definition.

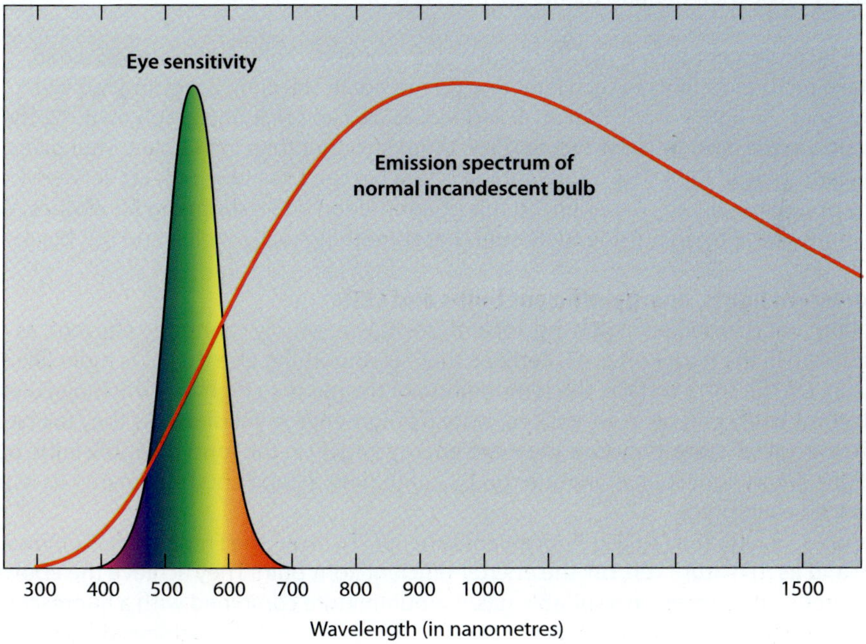

Eye sensitivity (left) and the emission spectrum of an incandescent light bulb (right). The small overlap illustrates that such lighting is quite inefficient.

Can we find a closer match for the two curves? An obvious solution would be to make the incandescent wire hotter than its usual 2,700°C, which would make the emission shift to the left. Recall that at higher temperatures not only is more radiation emitted, but the colour also changes. Think of an old-fashioned poker in a fireplace: its colour turns from red to orange to yellow as it heats up. Or think of a dimmed incandescent bulb. As we turn up the dimmer, it gives us not only more but also *'whiter'* light. The spectrum in the figure above thus shifts to the left, *i.e.*, toward the shorter wavelengths.

A filament wire that could be made *twice as hot*, in terms of its absolute temperature in kelvin, would be best. Then the peak would be at *half* its current wavelength, exactly where the human eye reaches maximum sensitivity. That, however, would require a temperature of about 5,700°C. That is approximately the temperature on the sun's surface. So, the solar radiation *does* peak at the eye's maximum sensitivity, which is no coincidence, of course.
Unfortunately, as we increase the temperature of the incandescent wire above its normal 2,700°C, we approach its melting point, ~ 3,400°C for tungsten. Raise the temperature and you reduce the lifetime of the incandescent bulb dramatically. Filling the bulb with a halogen gas helps to some extent. It allows us to raise the temperature somewhat, thereby increasing the efficiency by about 20%.
But the melting point of tungsten remains a solid limit. Other metals are no solution either because their melting points are even lower.

So, an incandescent bulb is not the ideal way to generate light. Furthermore, when we take pictures or films in such light and do not adapt the camera setting, we will notice that the light is not white at all. Indeed, as we can see in the graph on p. 43, the emission in the orange-red is two to three times stronger than in the blue. That gives us a warm type of light. The disadvantage is that we can't see blue objects very well. Many of us have already experienced this phenomenon while shopping for clothes. We sometimes have to go outside to be sure that something really is blue and not black.

Fluorescent lights, energy-efficient bulbs and LEDs
Lighting that is based on a glowing, incandescent wire simply cannot be efficient, as was shown in the figure on p. 43. Perhaps there is something else that acts more like the sun. On the sun's surface, the temperature of the gas is so high that the molecules and atoms easily end up in an 'excited' state by high-energy collisions. As they fall back to their 'original' state, they lose the extra energy largely in the form of visible light, our sunlight. Smart light sources imitate this by combining a good choice of materials with advanced electronics.
Examples include the familiar fluorescent tube (or 'TL', from the French *Tube Luminescent*) and its miniature version: the energy-efficient light bulb. They achieve the above light-producing process in a suitable gas/vapour mixture combined with a fluorescent layer inside the tube. The more recent *Light-Emitting Diodes* (LEDs) achieve this in a solid semiconductor.

The result is that the energy input is converted directly into radiation and not – as in the incandescent bulb – via the heating of the filament. Such modern light sources produce more light for the amount of energy used: more lumens per watt (abbreviated lm/W).

Efficiency (in lumens per watt) and life span of various light sources. Note the poor performance of the old-fashioned candle.

	Candle	Incand. light bulb*	Compact fluorescent 'saving lamp'	LED	Fluorescent tube (TL)
Efficiency (lm/W)	0.15	10–12	50–60	50–100	100
Life span (hours)	10	1,000	>5,000**	50,000	10,000

* Halogen bulbs achieve up to 14 lm/W and last up to 4,000 hours.

** Depends largely on the number of times one switches the light on and off.

How much better are these new light sources? They are all more efficient than the traditional incandescent bulb. Moreover, they last much longer because there is no hot wire that slowly but surely evaporates (which we witness as a blackening of the bulb's glass). The table gives an overview.

These values illustrate the potential efficiency one can achieve with LEDs. Depending on their colour they are already comparable to the efficiency of fluorescent tubes (TL). They also have the useful feature that they can produce light in a single colour: red, blue or green. That is particularly useful if a specific colour is desired, for example for brake lights or for traffic lights. It is, of course, more efficient to produce the desired colour directly instead of first producing the entire spectrum and then having to filter out a large part.

If we want to produce *white* light with LEDs, we can do that by simply combining the three colours. But we can also use an intermediate step, so that the LED's light first irradiates a fluorescent layer, which, in turn, emits 'white' light. This is very similar to the process involved in producing light in fluorescent tubes.

Furthermore, the LED offers possibilities that would be unimaginable with the old-fashioned light bulbs. For instance, one particular type of LED lamp that is already available can produce any desired colour. A suitable combination of a red, green and blue LED will yield every conceivable colour in the spectrum simply by separately controlling the electrical current in each LED.

If the light's colour is not that important, the efficiency of a light source can be extremely high. Energy-efficient street lighting using sodium lights reaches 200 lm/W. Just compare this to an old-fashioned candle with an efficiency of only 0.15 lm/W. If we were to replace one 100 W sodium lamp with candles to light the roads equally well, we would need 1,300 candles. (This would be like a big party along the side of the highway!)

What is the maximum efficiency that any one light source can achieve? This would require a situation where all of the light is radiated at the peak of the eye's sensitivity curve (550 nm, see figure on p. 43). In this case we would achieve 680 lm/W. However, it should be noted that not much would be left of the colour spectrum, since only yellow/green light would be emitted.

F | Comparison of Various Energy Users

Any honest comparison between various energy users should provide the consumption in terms of 'primary energy', *i.e.*, in terms of the amount of fuel needed to do the job. So, the conversion efficiencies of the power plants (most of which are powered by one or another fuel) or the efficiency of a car engine (that runs on petrol or diesel) etc. should also be taken into account.

The following table provides a rough comparison between the fuel consumption rates of various energy users. It provides the power in watts *and* energy use in terms of litres of oil or petrol (or cubic metres of natural gas) per hour – those are fairly comparable, after all. These are approximate values and they assume that the energy-using device is running continuously using the specified power.

Conversion efficiencies are also taken into account. For example, the energy a car uses in terms of power (watts) is calculated at five times the *mechanical* power of the engine, because engine efficiency averages 20%.

Energy Use Based on Primary Energy

	Power in watts	Energy use in m^3 of natural gas or litres of oil per hour
Electric clock (3 W)	10	0.001
Electric shaver (10 W)	30	0.003
Candle, small pilot flame, an adult human	100	0.01
Lamp, laptop (100 W)	300	0.03
PC, TV (200 W)	600	0.06
Medium-sized gas stove	2,000	0.2
Vacuum cleaner (1,000 W)	3,000	0.3
Kitchen geyser (natural gas)	10,000	1
Mid-sized central heating unit, shower	20,000	2
Car (100 km/h, 16 kW)	80,000	8
Full car, per person	20,000	2
Full aircraft, per seat	400,000	40

Some interesting conclusions can be drawn from this table. For example, the energy used by an old-fashioned pilot flame or a candle in a single day is enough to power a TV for an evening, or take a 10-minute shower. The energy used by a 100-W electrical appliance for a day is enough to drive a car for 10 kilometres. And a full aircraft uses 20 times as much fuel as a full car on the highway, per passenger, per hour. But, since the aircraft flies roughly ten times as far in the same period of time, it uses only twice as much fuel as a car per kilometre and per person. So if 2 people take a car to go on vacation, they use about as much fuel per person as a charter flight with no empty seats.

G | Energy Needed to Produce Something

When we get into our car or switch on the TV at home, we can usually get a fair idea of how much energy we're using. But that is not the whole story. There is a hidden figure in our energy-use bookkeeping: the energy used in the manufacturing process. How much energy does it take, for example, to make a car, a TV, an electric razor or to get a kg of cheese on the table? This question is not easy to answer. What exactly should we take into account? Does it include, for example, the petrol used by people to get to work? (Perhaps they would have used their car anyway, to drive to the beach, for example.) And should we also take into account the energy that is saved, or used, when recycling or discarding a product?

There is a whole new science emerging to study this aspect of energy use: life-cycle analysis. The 'cradle-to-cradle' concept is an exponent of the effort to use as little energy and material as possible in the entire life-cycle of a product.

How much energy does it take to produce:	in MJ	in litres of oil
1 kg of steel (from ore) [1]	15	0.4
1 kg of aluminium (from ore) [2]	140	3.5
1 kg of aluminium (recycled) [2]	7	0.2
1 kg of paper [3]	20	0.5
1 kg of glass [3]	12	0.3
1 kg of wheat, at farm [3]	3	0.08
1 kg of soy beans, at farm [3]	1.3	0.03
1 kg of potatoes, at farm [3]	0.7	0.02
1 kg of cheese [3]	34	0.85
1 kg of sheep, for slaughtering, at farm [3]	7.1	0.18
1 car [3]	82,000	2,000
large domestic appliances, per kg [4]	20	0.5

1. Corus NL (2008): total primary energy, including coal, is 15,6 MJ/kg
2. US Department of Energy (www.eia.doe.gov)
3. Ecoinvent database (www.ecoinvent.ch)
4. Traditional rule of thumb

The figures in the table on p. 47 give some indication of energy use. As already noted above, these figures should be used with caution. Most of the figures tend to decrease over time as efficiency increases. The data provided by Ecoinvent refer to 'cradle to gate', *i.e.*, energy use from raw material to product leaving the factory or farm. Note the importance of recycling aluminium: mining and production of aluminium takes about 20 times as much energy as recycling.

Conclusions regarding Domestic Energy

- In a moderate climate, home heating is often the largest energy user. Home insulation and energy-efficient windows are therefore essential. Even tightly drawn curtains make a big difference.
- If, during the heating season, the average temperature difference between inside and out is 14 degrees Celsius (like it is in the Netherlands), lowering the thermostat by 1 degree lowers the heating cost by 1/14 or 7%.
- Heating water requires a lot of energy. A running hot-water tap uses as much energy as a few hundred or even a thousand lights left on for the same period of time. So it pays to use the shower and the (dish)washer wisely.
- A typical Western European household consumes 400 W of electricity on a year-round day-and-night average.
- Appliances that stay cool during operation do not consume that much electricity. We can judge from the heat production whether or not an appliance is a big electricity consumer.
- An electric boiler is almost certainly the leading consumer of electricity.
- The washing machine, dishwasher, dryer, refrigerator and freezer each consume at least as much as the total lighting for an average home.
- It is smart to place the refrigerator in a relatively cool place and set the thermostat at a temperature not colder than necessary.
- All electric appliances, including lights, release their total electricity consumption as heat into the home (except for those that dump their heat outside the house, such as the (dish)washer).

5 | Moving Around: Energy and Transportation

It may have gone unnoticed by many of us, but moving around has probably become the leading item on our personal energy bills. This is not only because petrol is heavily taxed: also in terms of energy, the average family car may be the chief private energy user in many developed countries. For example, if a family drives an average of 20,000 km annually, it consumes circa 1,400 litres of petrol. In terms of energy use, this is about the same as the 1,600 m^3 of natural gas consumed annually by the average Dutch household.

So let us look at *how* we travel. In most developed countries, the average resident uses the car as his or her primary mode of transport. In the Netherlands, for example, the average person travels a total of 32 km per day, of which 25 km is by car, 3.6 km by public transport and 2.5 km by bicycle (data Statistics Netherlands, 2007). In rapidly developing countries like China, the use of cars may still be moderate, but car sales are soaring.

It is, therefore, a good idea to look first at the car. How efficient is it? Can we improve it? If so, how?

A | Cars

In most countries, the largest proportion of car travel is on horizontal roads. This means that it does not involve much climbing and descending, and, since the two compensate one another to a large extent anyway, let us focus on flat roads. First, we obviously need energy when we accelerate. The energy needed to reach a certain speed is proportional to the mass of the car and the speed squared ($\frac{1}{2} mv^2$, to be precise). This means that accelerating to 100 km/h takes four times as much energy as accelerating to 50 km/h. This is good to keep in mind whenever we notice the traffic light turn yellow in the distance. If we proceed carefully we may be able to avoid braking and having to accelerate again. Anticipation is the keyword here.

How much fuel does an extra complete halt take? If we do the calculations we discover that, roughly speaking, accelerating from 0 to 100 km/h takes as much fuel as driving 1 km on a highway. Incidentally, for a train this would be about 15 km. This is because a train is relatively heavy and is pretty economical once in motion. The lesson is that trains perform well provided that they do not stop too frequently, unless they recover the braking energy.

Since petrol (or diesel, or whatever energy resource) is consumed primarily on the highway, let us look at this situation and find out what exactly determines energy use. When we drive at constant speed, our car has to cope with a certain amount of resistance. In fact, the engine has to push the car with exactly the same force that is needed to overcome that resistance. And it is precisely the resistance that determines the fuel

consumption, as we will see later.
What *is* that resistance we have to cope with? There are two forms of resistance, and we will discuss them separately.

A1 | Rolling resistance

The ball bearings do not really play much of a role when it comes to the *rolling resistance* of a car. Rolling resistance is almost entirely caused by the deformation of the tires on the road. That may seem odd: Isn't rubber elastic, and doesn't it bounce back nicely into its original form after it is compressed?

That is indeed true, but the process of compression and expansion does take some energy. This is because the force necessary to compress rubber is greater than the 'push-back force' that the tire delivers when it expands (see box 'Rolling resistance and air resistance', p. 51). The difference between the two – the net energy – is converted into heat. You can feel how warm the tires get after you drive on a highway for a while! The resulting resistance from all this – the rolling resistance, or rolling friction – is proportional to the weight of the car. It is, in fact, a fixed percentage of that weight, given a certain type of tires at a certain tire pressure. That percentage is called C_r, the 'rolling resistance coefficient'.

What *is* the value of this C_r? Could it be 1 or is it 0.1? These high values can't be right, and here is why. Suppose our car is parked on a slope, and we release the brakes. How much of a slope do we need for the car to start rolling? It turns out that a slope of about 1% is enough. Now we have to find the force that pushes the car forward on such a slope. Let us look at the weight of the car (which points in the vertical direction) on this sloping road. What we need is its component parallel to the road. We find that this component is 1% of the car's weight in this case. *That* is the force which pushes the car forward. So, if that is enough to overcome its rolling resistance, the value of C_r must be about 0.01. And sure enough, it is – at least if the tire pressure is right. It is only logical that C_r increases as the tire pressure decreases, because the whole story boils down to the deformation of the tire. It turns out that the rolling resistance increases by about 5% if the pressure is decreased by 10% (*Tires and Passenger Vehicle Fuel Economy*, National Research Council of the National Academies, Washington, DC, 2006).

The value of C_r varies for different tires and road surfaces. In this respect concrete performs best. It does not give way at all and consequently has a somewhat lower resistance than asphalt, which is just a bit softer. For the average car, a reasonably precise value for C_r is 0.010. But, for modern tires, it is already a bit lower at 0.009. Future figures of 0.006 seem plausible. It is important to note that speed does not play a role here. The rolling resistance is, for all practical purposes, independent of speed.

A2 | Air resistance ('drag')

The air resistance, or drag, behaves differently. It goes up with the square of the speed: if we double our speed, the air resistance will be four times as large. We could guess that from experience: if we stick our hand out the window at a speed of 100 km/h, the force we feel is far larger than at 50 km/h. It is four times as large, to be precise. Furthermore, the air resistance is proportional to the density of the air. This means that the air resistance increases with increasing atmospheric pressure. It also means that, in the heat of the summer, the air resistance is lower than in the winter, given a certain atmospheric pressure. This is obvious when we realise that warm air expands and thus gets 'thinner'. Analogously, also the thinner air at higher altitudes reduces drag. This is the same for bicycles and speed skaters: breaking a speed record is easier at higher altitudes.

Another factor that determines the drag is the size of the car, or, more precisely, the frontal area. Low and narrow cars have less air resistance than high and wide ones.

Finally, the air resistance is proportional to the drag coefficient C_d, which is determined by how streamlined the car is. While the value of C_d can be as high as 1 for clumsy objects that are not streamlined at all, it can go down to 0.25 for some late-model cars.

Why are rolling resistance and air resistance so important? It is because they have a direct influence on fuel consumption. Together, they comprise the *total* resistance that a car undergoes when it is driven along a horizontal road. This is equal to the force that the engine must continuously produce to maintain a constant speed.

Rolling resistance and Air resistance

The rolling resistance F_r is written as

$$F_r = C_r \times mg,$$

with C_r being the rolling resistance coefficient, m the mass and g the acceleration due to gravity. This resistance is mainly due to the fact that the force needed to compress rubber is larger than the force which is delivered by the rubber as it tries to bounce back to its original form (in terms of physics: the closed integral $F \cdot dS$ is *not* equal to zero). The energy 'lost' in this process is released as heat.

The air resistance or 'drag' F_d is proportional to the square of the speed v. We may derive directly from Bernoulli's law ($p + \frac{1}{2}\rho v^2 = \text{constant}$, where p is the pressure and ρ the mass density of air) that the pressure right in front of a flat plate, where $v = 0$, must be equal to $\frac{1}{2}\rho v^2$. Consequently, the *force* on such a plate is $A \times \frac{1}{2}\rho v^2$ where A is the area of the plate. However, most objects have some level of streamlining, which reduces that force, or resistance. This is expressed in the 'drag coefficient' C_d. So, the air resistance on an object with a frontal area of A becomes

$$F_d = C_d \times A \times \frac{1}{2}\rho v^2.$$

The value of C_d is around 0.25 for late-model cars. The lower limit for the ideally streamlined vehicle, which basically corresponds to the shape of a fish or a teardrop, is 0.05.

Although the value of C_d is not quite independent of speed, we may consider it as a constant for the relevant speed range.

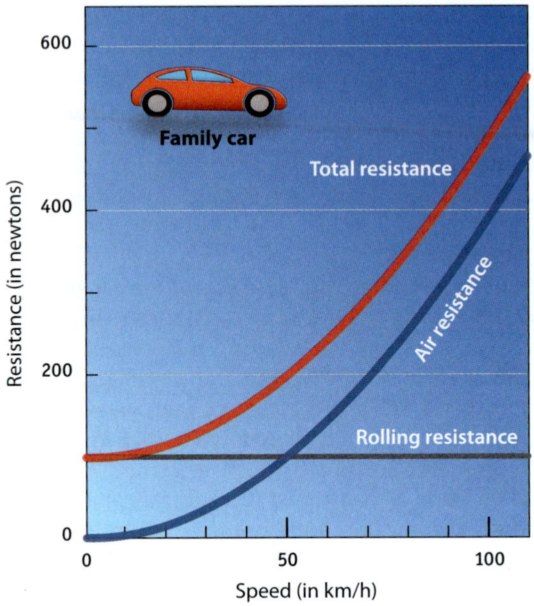

The resistance that a car experiences increases dramatically at higher speeds, because of the rapidly increasing air resistance or drag (blue curve). This increases fuel consumption at higher speeds. The figure corresponds to a car with a mass of 1,000 kg, a rolling resistance coefficient of $C_r = 0.01$, a drag coefficient $C_d = 0.4$ and a frontal area $A = 2\ m^2$.

The interesting point is that this force is equivalent to the work done – or the *energy* used – *per unit distance*. And this can be directly translated into a car's energy consumption (see box 'From resistance to fuel').

To assess the influence of the two types of resistance on the fuel consumption separately, we need to know their relative magnitude. The two are shown for a typical mid-sized car in the figure above. We notice that, when we drive through residential areas at lower speeds (below 50 km/h), the rolling resistance is dominant. It is clearly an advantage in this situation to have a light car (especially since it uses less energy during the frequent stop-and-go cycles). Things change on the highway, however. Here air resistance is dominant. Now it is more important to have a small car in terms of its frontal area: low and narrow. Moreover, streamlining is essential here. So, driving with your windows open or carrying stuff on the roof is not particularly efficient in that it can increase fuel consumption dramatically. Since the air resistance increases with the speed squared, driving fast *does* consume more fuel.
Fortunately, things are not as bad as suggested by the steeply rising curve. This is because the efficiency of an engine increases at higher speeds, since it has to work harder (the efficiency is zero when the engine is idling!). This effect partly compensates for the steep increase in air resistance. But driving fast, in general, remains inefficient. Therefore, some sort of speed limit may be a good idea after all, if we consider energy use.

From resistance to fuel

The graph on page 52 shows the behaviour of rolling resistance and air resistance (or drag) as speed increases, for a relatively light and poorly streamlined car. The equations for the two kinds of resistance are provided in the box 'Rolling resistance and Air resistance' on page 51. The values used in this figure are: mass 1,000 kg; rolling resistance coefficient $C_r = 0.01$; drag coefficient $C_d = 0.4$; frontal area $A = 2\ m^2$.
We notice that for this car, the two types of resistance are equal at 50 km/h. For a heavier and/or more-streamlined car, the two curves cross at a higher speed. Incidentally, for a cyclist, that crossing point is at circa 15 km/h.

Note that the unit on the vertical axis is the unit of force: newton. This is a direct measure of energy consumption, which is easy to see. Given *Work = Force times Distance*, it follows that *Force = Work per unit distance, i.e.*, 1 N = 1 J/m. At 100 km/h for example, we find from the graph that for this car it is 500 newtons, or 500 joules per metre, or 500 kilojoules per km.
But this is the *mechanical* energy. At an estimated engine efficiency of 20%, this translates into a *total* energy use of 2,500 kJ/km. Because oil and petrol have a combustion value of about 40,000 kJ per kg, this amounts to 6.3 kg per 100 km, or a fuel consumption rate of 7 litres per 100 km.

From resistance to fuel

An experimental illustration of how fuel consumption is dependent on speed is shown in the figure below. With decreasing speed and while driving in fifth gear, an efficiency of 29 km/l of petrol was achieved, compared to 20 km/l at 90 km/h.

Obviously, the results will vary for different cars. But the trend remains the same: it pays off to watch our speed. Moreover, we must aim for high engine efficiency. In this case, this was achieved by using fifth gear throughout, even at 40 km/h.

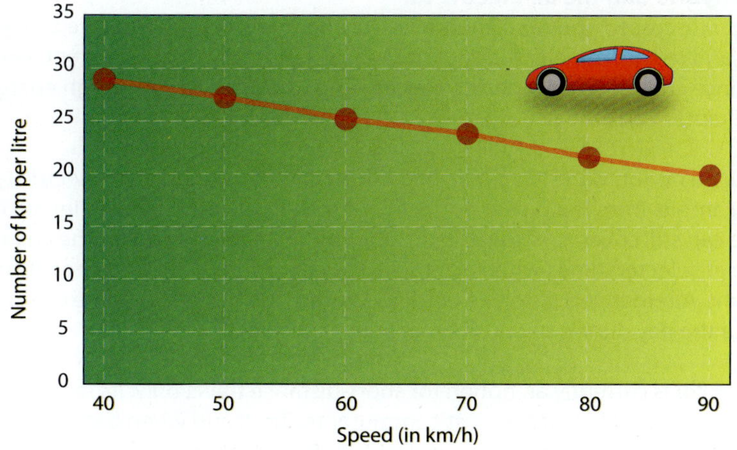

Fuel efficiency in terms of number of kilometres per litre for speeds between 40 and 90 km/h in fifth gear (Toyota Yarns, courtesy of Herbert Blankesteijn).

The hybrid car

It does not sound particularly logical that in order to increase efficiency we equip a car with two engines: an ordinary internal combustion engine and an electric motor. Yet this is indeed the case with a hybrid car. It derives its energy exclusively from petrol or diesel, so all it does is use this energy more efficiently than an ordinary car. How does it achieve this?

The secret is that its electric motor – in combination with a relatively large battery – makes driving in the city, with its frequent stop-and-go cycles, much more efficient than an ordinary car. At a traffic light, the motor simply turns off; it does not idle like the engine of an ordinary car. Furthermore, it recovers its braking energy by converting it into electricity that recharges the battery. That energy is lost in the form of heat in the brakes of an ordinary car. So, all of a sudden, driving in the city has become much more efficient. This may not be surprising if we recall that the air resistance (or drag) is small at low speeds. All that is left to conquer is the rolling resistance. Thus, the total resistance is small, allowing for greater driving efficiency.
On the highway, this advantage disappears. Moreover, the battery only has a relatively short range and the car must revert to running on petrol or diesel, just like a 'normal' car. This means that, for long-distance travel, a 'normal' car with the same amount of streamlining is just as efficient.

Since the electric motor may be used in conjunction with a normal engine for a short time, to go up a hill or overtake another car, it does not hurt to have a smaller internal combustion engine than you would normally find in such a car. This can further enhance engine efficiency, since the efficiency increases when the engine 'has to work harder'.

In conclusion, the advantage of a hybrid car is mainly evident in the city. So, it may be a good idea for taxis to switch to hybrid cars. Not only would that make driving more economical, it would also reduce air pollution in the city.

The plug-in hybrid and the all-electric car

The hybrid car discussed above still gets all of its energy from fossil fuel. We can go a step further and *really* drive electrically. For long-distance travel this is not yet a practical option, since current battery performances keep us from having enough energy on board (see chapter 12).

An intermediate option is the *'plug-in hybrid'*. This enables us to drive on electricity drawn, for example, from the power outlet in our own homes. If the batteries are running low, we can still cover a reasonable distance by switching to petrol. The charging can be done at selected times when electricity consumption is low (and cheap), for example at night. Alternatively, if solar electricity is used, the batteries can be charged in the middle of the day during periods of maximum sunshine.

The *all-electric car* is currently an option for short-distance travel only. Its range (typically 150 km) may be sufficient for most commuter traffic. If and when battery development improves, cars may drive electrically for distances comparable to those commonly covered by current combustion-engine cars.

Obviously, the question arises whether the electric car is more energy-efficient than the traditional car running on fossil fuels. To compare the two, we must compare the efficiency of an internal combustion car engine (20% on average) with electricity generation. If we use a power plant fuelled by oil or natural gas, a net efficiency of 35% is a reasonable figure. So, if losses in the battery-charging process and in the electric motor are small, the electric car would prevail. The advantage is even larger when we consider the lighter weight of the electric motor compared to the internal-combustion engine, plus the fact that an electric car is more efficient in city traffic with its many stop-and-go cycles, since it consumes zero energy when it is standing still. However, energy losses in the charge/discharge cycles can largely offset these advantages.

Electric cars may also be powered by electricity from fuel cells, which are addressed in chapter 12.

B | Buses and Trains

Buses and cars both use rubber tires. Therefore, the rolling resistance *coefficients* should be similar, but the rolling resistance itself is for a bus much larger because of its weight. For a fair comparison, we should take the rolling resistance *per seat*. If we do that, we discover that cars and buses are quite comparable (see table on p. 56). In other words, if we assume that both have a 100% passenger load, cars and buses should have almost the same energy use per person at low speeds.

However, things change at higher speeds, when air resistance, or drag, is more prevalent. Granted, the bus has a larger frontal area and less streamlining than most cars, but this is compensated for by the large number of passengers a bus can transport. The result is that a bus has only half the drag per passenger compared to a car. So, at higher speeds the bus would beat a car by a factor of 2, as observed in the table on p. 56.

Experimental data for buses

An average bus for 50 passengers consumes about 1 litre of diesel every 3.5 km. This amounts to 175 passenger-kilometres per litre (the number of passenger-kilometres is the number of passengers multiplied by the number of km travelled). In contrast, a typical diesel car with 4 people on board and travelling at the maximum bus speed limit may reach 80 passenger-kilometres per litre. This means that the car is only half as efficient as a long-distance bus.

Trains are a lot like buses if we consider total *air resistance*. However, trains carry a much larger number of passengers than buses do. So, if we consider the air resistance *per seat*, the train gains a large advantage (see table on p. 56). It beats buses by a factor of 2 and cars by a factor of 10.

The approximate values for the quantities that determine the energy use of cars, buses and trains. The figures for rolling resistance and air resistance are calculated per seat. These numbers thus provide a direct comparison, if we assume the same relative passenger load (see also table on p. 60).

	Mass per seat (kg)	C_r	**Rolling resistance compared to a car**	Frontal area per seat (m^2)	C_w	**Air resistance compared to a car**
Car	300	0.01		0.5	0.4	
Bus	300	0.01	**1**	0.15	0.7	**0.5**
Train (double-decker)	600	0.0015	**0.35**	0.025	0.8	**0.1**

Another big difference is revealed in the values for *rolling resistance*. Trains have steel wheels and run on steel rails. Therefore, the rolling resistance coefficient is much smaller. It may not be quite as low as one would expect from the stiffness of steel. This is caused by non-ideal wheel-rail contact that causes some extra friction, and/or by poor track maintenance. Reasonable figures for C_r range from 0.0015 to 0.0020. This value is still much lower than figures for rubber tires on roads. This advantage is partially offset by the relatively large weight of a train, but the rolling resistance of trains per seat remains about three times lower than for busses and cars (see table above).

So, trains beat buses and cars especially at speeds at which air resistance is greater than rolling resistance. For trains, the two resistances are equal at around 70 to 100 km/h, depending on the type of train. Remember that for cars this happens at a somewhat lower speed, 50 to 70 km/h, again depending on the type of vehicle.

Stopping and accelerating has not yet been taken into account. Trains are relatively heavy, so accelerating is rather costly. Since cruising is relatively efficient, a train can cruise for quite a distance and use the same energy as needed for accelerating to cruising speed. If we work it out for the train used above in the table, which consists of four double-decker cars, we find that this distance is 15 km. For a car, this figure is only 1 km. The conclusion here is that trains are especially efficient at longer distances. The more they stop the less efficient they become, unless the braking energy is recovered (a technique that is being used more and more frequently).

During the winter, trains have to be heated, just like cars. But, unlike cars powered by internal combustion engines, they do not have the waste heat to recycle. Instead, they use electric heating, which can substantially reduce a train's energy efficiency.

Experimental data for trains
A double-decker train consisting of 4 cars (372 seats in total, 254 tons) turns out to have an energy-efficiency of 158 passenger-km per litre of petrol, if converted to electricity. A 100% passenger load and an average of 14 km between stops has been assumed. This is about three times as efficient as a full car, which reaches about 56 km/ litre, assuming four occupants and an average of 14 km per litre.

The advantage of trains increases at longer distances between stops. For 50 km between stops, the energy-efficiency increases to 260 passenger-km per litre. Again, it is assumed that braking energy is simply lost.

TGV (*train à grande vitesse*), **High-speed** trains, **MagLev** (magnetic levitation) all move at very high speeds through air at ambient pressure, so they are bound to be inefficient. The reason is that the air resistance increases with the speed squared, just like for cars. So, twice as fast means four times the energy.
The rolling resistance becomes negligible at speeds above approximately 300 km/h. Therefore, magnetic-levitation trains (which have no rolling resistance) offer barely any improvement in terms of energy efficiency.

So, one would not expect high-speed rail travel to be very efficient. Despite its excellent streamlining (low C_d) we should expect that a TGV, travelling more than twice as fast as a normal long-distance train, will use about four times the energy.

Experimental data for TGV
In practice, the TGV uses roughly five times as much energy as an ordinary long-distance train if this is expressed in terms of petrol consumed (as we did with the trains above). It achieves about 50 passenger-km per litre. This confirms our calculations above, where we expected it to be roughly a factor of 4 worse than a regular long-distance train. One aspect that may improve TGV performance in practice is the relatively high occupancy.

C | Aircraft

The fuel consumption of planes per passenger-km depends on a number of variables. Is the plane full? What type of plane is it? How close are the seats to one another? Is the plane carrying cargo in addition to passengers? Is it a long-distance flight?

The last question has an interesting answer. Obviously, short flights are not very efficient, because of the relatively large amounts of fuel used for take-off and landing. So, longer flights should be more efficient. But wait, very long flights require extra heavy loads of fuel. Indeed, flights above about 5,000 km become *less* efficient, since there are fewer passengers on board because of the weight of the extra fuel. We can estimate this by using a rule of thumb that an aircraft at cruising altitude consumes almost 10 cm^3 of fuel per second per seat. For a flight of 5,000 km that takes 6 hours, we can expect fuel consumption at $6 \times 3{,}600 \times 0.01$ litres per passenger. This is 216 litres, or the equivalent of the weight of two passengers!

To estimate fuel consumption, let's again use the rule of thumb that the fuel consumption for each passenger is almost 10 cm^3 per second. This number was fairly accurate for many different types of planes during the 1980s, but it has been gradually decreasing somewhat as planes are becoming more efficient. Even so, it is convenient to keep this in mind as a rule of thumb.
It follows that flying for one hour consumes 36 litres of kerosene per passenger. Since we cover almost 1,000 km during an hour, the consumption should be about 3.6 litre per 100 km, or 28 km per litre per passenger. This is equivalent to the fuel efficiency of a car consuming 14 km per litre with two occupants.

The conclusion is that a full plane is (per km and per passenger) roughly as efficient as a half-full car. Since both planes and cars are becoming increasingly efficient, this rule of thumb may hold for many years to come.

There are some factors that complicate matter, however. Take cargo. Most planes carry a lot more cargo than just passengers' luggage. This makes it difficult to calculate efficiency with any precision (see also experimental data for aircraft below).

It may come as a surprise that planes are relatively efficient, given the fact that the air resistance increases steeply with speed. Why is that? We should bear in mind, first, that the cruising altitude for planes is at (or even above) 10 km. The air density is only one-quarter of the air at sea level. Furthermore, planes have excellent streamlining. Last but not least, their dead weight is relatively small. If we compare the total mass of passengers plus luggage plus cargo with an empty aircraft, we find a ratio of about 1 to 10. This is particularly favourable if compared with a passenger ship (see p. 60).

Experimental data for aircraft

A long-distance flight by a Boeing 777-200, carrying 260 passengers and 12,000 kg of cargo, consumes about 68,000 kg or 85,000 litres of kerosene over a distance of 9,000 km. This is 28 passenger-km per litre, which corresponds to the above figures. But we should also consider the cargo. If we assume the above 12,000 kg of cargo to be the equivalent of 120 passengers including their luggage, all of a sudden we have 380 people on board. This would give us 40 passenger-km per litre.

Comparison

How do various means of transportation compare, if we consider energy use? The table on page 60 quotes some approximate values.

What about the Zeppelin?

An airship, or Zeppelin (named after the German count Ferdinand Graf von Zeppelin, the inventor of the airship) can stay up in the air without using any energy. This sounds extremely efficient. So doesn't this offer the ultimate energy-efficient form of transport?
Well, as long as the Zeppelin is just hanging around in the air, this may be true. Even if it is used as a crane, to hoist things and transport them over short distances, it may be an efficient piece of equipment. But as soon as it starts moving, this all changes.
For example, could a Zeppelin beat a plane on a trip across the Atlantic, if we look at energy use?

Let's make a back-of-the envelope calculation. We only need to consider air resistance here. This increases with the speed squared. A speed of 100 km/h seems to be a minimum (just think of headwind!). At that speed, the air resistance is gigantic for such a large object moving through the air. We have to assume air density at approximately sea level, because the airship would not have enough lift at high altitude where the air pressure is lower.
We may take the dimensions of the Hindenburg, the airship that made quite some history when it tried to land in New Jersey, USA, back in 1937. It was carrying 100 passengers and had a diameter of 41 metres. This yields a frontal area of 13 m^2 per passenger.
Obviously, there is no way that this can compete even with a car, which has a frontal area per passenger of only 0.5 m^2. Even if we include the (relatively small) contribution of the rolling resistance of a car and the superior streamlining of an airship, the car would still be about 10 times more efficient than the airship.

We can check our calculations using the Hindenburg's technical data. It had a top speed of 135 km/h and its engines had 3,560 kW of total power. If we work it out we discover that, indeed, the car still beats the airship by a factor of 8.
If compared to a plane, the airship has two drawbacks. First, it has this enormous volume, which creates large amounts of air resistance. Second, the density of the air through which it moves is larger by a factor of 4 compared with the air at the cruising altitude of a plane.

The conclusion is inevitable. There is no bright future for the airship (unless we *really* want to take our time, of course).

Energy efficiency in terms of number of passenger-km per litre of fuel, assuming a passenger load of 100% and long-distance travel. For electric vehicles, the electricity is assumed to be generated in an oil-fuelled power plant, and for the bicycle the energy equivalent of food is expressed in litres of oil.

	Number of passengers	Speed (km/h)	Energy efficiency (passenger-km/litre)
Bicycle	1	20	500
Electric bicycle	1	20	400
Train	250	130	250
Bus	50	100	170
Car	4	100	60
TGV	377	300	50
Aircraft	400	900	30*
Passenger ship	2,000	50	4

* Modern planes like the A380 may achieve 35 (Source: Singapore Airlines, 2008).

The poor performance of the passenger ship is striking, especially in view of its low speed. The reason is not only that it has to move through water, the density of which is three times greater than air at sea level. More importantly, the dead weight of a passenger ship – or a cruise ship – is enormous. Unlike other means of transportation, these ships have been designed to spend a lot of time at sea. Therefore, they have dining rooms, shops, ballrooms, even swimming pools and the like. In fact, these ships are floating villages. For example, take the relatively modern Queen Elisabeth 2. This ship is 345 metres long, has a mass of 70,000 tons and carries 1,900 passengers. If we include their baggage the passengers weigh 200 tons. This is only about 0.3% of the ship's total weight. The passengers on an aircraft comprise 10% of its total weight, as mentioned above.

Care should be taken not to generalise the poor performance of passenger ships to ships in general. The load carried by cargo-boats and oil tankers is usually very large compared to their own weight. Therefore, their energy efficiency may be very good.

D | The Bicycle

Mankind managed to improve its transport efficiency by a factor of four by inventing the bicycle. Walking is already fairly efficient compared to flying and other modes of transport, including those in the animal world. But cycling beats every mode of transport, if one measures the amount of energy per kilogram of displaced mass per unit distance, as is shown in the figure below.

Why is bicycling so much more efficient than walking? It is partially because the centre of gravity of a cyclist remains constant: when one leg rises the other descends. This is not the case with walking: our centre of gravity moves up and down about 3 centimetres with every step, which obviously requires some energy. Walking is mechanically complex, involves many different muscles and, in the end, makes it much less efficient than bicycling.

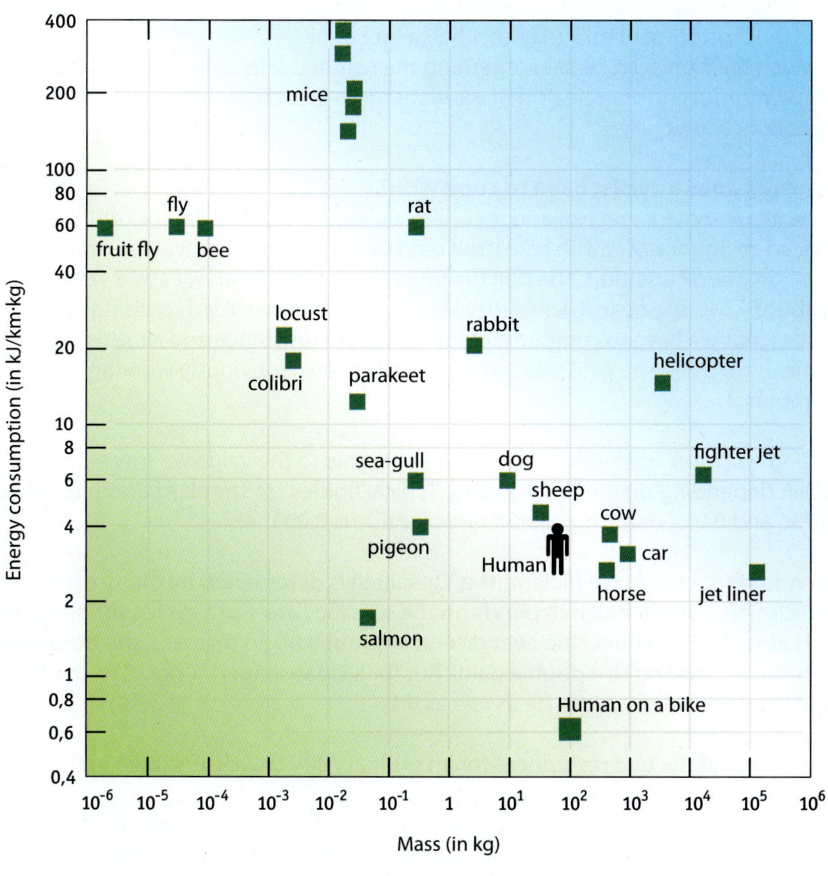

Comparison of the energy used per kilometre and per kilogram transported mass, arranged from light (left) to heavy (right) (data from S. S. Wilson, 'Bicycle Technology', Scientific American, March 1973, p. 90)

Just how much energy does it take to ride a bicycle? There is an easy way to calculate this. The power that it takes to ride a bike is quite comparable to climbing stairs, which we figured requires 100 W (see box 'Climbing the stairs', p. 20). Now imagine the following experiment in which we ride a bicycle continuously for 24 hours. We have already noted that 100 W of mechanical energy means that *total* energy use for cycling is 400 W, given a muscle efficiency of 25%. We also saw that a continuous energy use of 400 W means one litre of oil per day (see p. 19). So, if we would cycle for a day, we would use the equivalent of 1 litre of oil for a 24-hour bike ride. In those 24 hours, we cover a distance of 24 × 20 km, approximately 500 km.
Conclusion: A cyclist consumes roughly '1 litre per 500 km'.

That is enormously efficient. But there is a catch. We don't cycle on oil or petrol, but on bread, milk, steaks, and cheese. All of this food takes a lot more energy to produce and deliver to our home than its own energy content represents. Take milk, for example. The cows have to eat, they have to be milked (electrically), and the milk must be cooled down, transported, heated to be pasteurized, cooled again, and transported again. If you add it up, one litre of milk may cost half a litre of oil before it gets to your kitchen table. Much the same can be said regarding the rest of the food as well. So, while bicycling is wonderful and very healthy, if we want to save energy we might be better off buying a light motorcycle.

What forces does a cyclist have to cope with?
In places like Denmark and Holland it's relatively easy. These are flat countries, so we don't need to fight gravity. The bike itself offers very little mechanical resistance, provided it is in good condition. The ball bearings cause a loss of about 1%, a well-oiled chain about 1.5%, the gears (*'derailleur'*) 5% at most. Once we reach cruising speed, the real forces that we have to compensate for are rolling resistance and air resistance. It is the same as for a car (see p. 51). But with a bicycle, we are obviously working with different numbers.

The *rolling resistance* coefficient C_r is very comparable to the value we saw earlier for a car, again depending on the tire pressure. Typical figures for standard tires are between $C_r = 0.006$ and 0.01. For racing bikes a typical value is 0.003.

The *air resistance* or *drag* coefficient (the 'C_d-value') is determined by the streamlining. For bicycles, this pretty much depends on the specific case. For a cyclist sitting upright the C_d is about 1. For a hunched-over racer it is about 0.88 (in this case, the frontal area is also reduced, leading to a double gain). But for a cigar-shaped super bike, the HPV (*Human Powered Vehicle)*, it can be as low as 0.1.

The behaviour of the two resistances for an ordinary city bicycle is shown in the figure on p. 63. We see that the rolling resistance dominates for speeds up to about 15 km/h. At higher speeds, the air resistance is dominant. For reaching high speeds, streamlining is essential, and setting a speed record essentially means reducing air resistance. Super bikes (HPVs) can reach speeds over 100 km/h, which an ordinary racing cyclist cannot even think of.

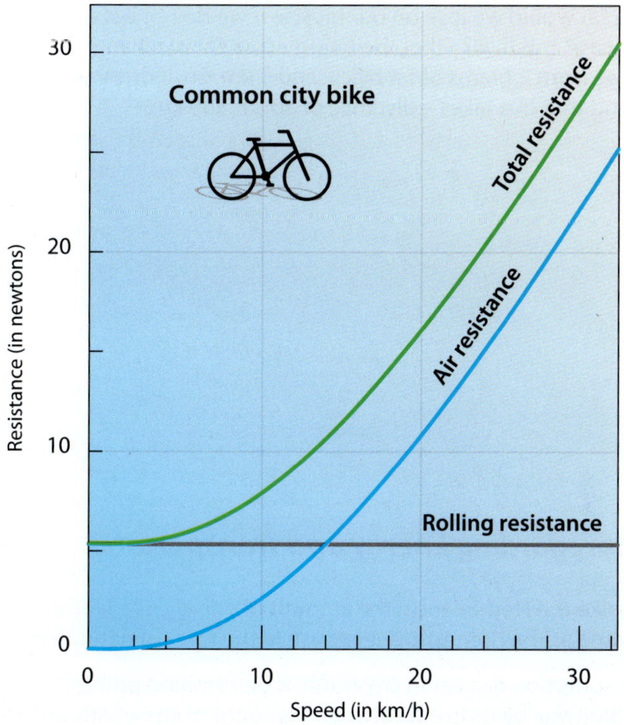

The resistance of a standard bicycle is dominated by air resistance (or drag, blue curve) above circa 15 km/h.

The figure contains another interesting detail, which we also noticed in the car case (p. 53). The total resistance, or force, is a direct measure of energy use per kilometre. When we see a force of 20 newtons in the figure, this is the equivalent of 20 joules per metre, or 20 kJ per km. As speed increases, the total force increases, as does energy use per kilometre. Consequently, cycling fast not only costs more energy per hour (which is obvious because we have to work harder), but also more per kilometre. So, if we cover a specific distance on a bicycle, and we do this as a work-out, cycling fast is indeed more effective than cycling slowly.

How much is the energy use if we take the case shown in the graph? This can be readily calculated as follows.
At a speed of 20 km/h, the total resistance for an ordinary city bike ($C_r = 0.006$; $m = 90$ kg; $C_d = 1.1$; $A = 0.50\ m^2$) is about 16 newtons. Mechanical energy use therefore is 16 kJ/km. With a muscle efficiency of 25%, this means a total energy use of 64 kJ/km.
Since the energy content of oil or petrol is 40 MJ/kg, this corresponds to 1.6 gram per km. So our earlier rough estimate of 1 litre per 500 km, *i.e.*, 2 grams per km (p. 62) was fairly accurate after all.

And how much fat would we lose on our bicycle if we do not eat anything extra? If we take the above case and realise that the energy content of oil and fat are quite close, we conclude that we burn 2 grams of fat by cycling 1 km. So, if one wants to lose 1 kg by bicycling, we find that this takes a distance of about 500 km.

Conclusions regarding Energy and Transportation

- In a country like the Netherlands, the amount of energy used for transportation is about the same as the domestic energy use in the form of natural gas.
- The fuel consumption of a car in city traffic is determined primarily by its weight, and on the highway by its frontal area and amount of streamlining. Driving fast is inefficient.
- Hybrid cars are relatively efficient, but primarily useful in the city.
- Electric cars are somewhat more efficient than petrol or diesel cars. However, battery performance and battery cost have so far limited their use to relatively short distances such as commuting within an urban area.
- For long-distance travel, a full bus is three times more efficient than a car. When frequent stops are made in a city, however, buses and cars are fairly comparable.
- For long-distance travel, full trains perform better than buses by about a factor of 4 in terms of energy use per passenger. This advantage vanishes rapidly if stops are frequent, unless braking energy is recovered.
- High-speed trains like the TGV are much less efficient than standard long-distance trains, because of high drag at high speed. Their performance is roughly equal to that of a car.
- If calculated per seat, a full plane uses about as much energy per km as a car carrying two people.
- The airship ('Zeppelin') is efficient only if it travels very slowly.
- The bicycle beats them all, reaching about 500 km per litre if the energy content of the food is expressed in terms of petrol consumption (not counting the energy needed to produce the food).

6 | Energy National and World-wide

A | Differences Around the Globe

Energy use varies tremendously for different countries around the globe. This is the case both for electricity and for *total* energy use. A few examples are given in the table below. Here, energy use per capita is expressed in kilowatts on an average continuous basis, rather than in one of the many exotic units often used, like tons of oil equivalent per capita per year.
The third column reveals that the two extremes differ by almost a factor of 50 in total energy use per capita, and even more if we consider electricity as shown in the fourth column.

Energy use and gross domestic product per capita in various countries in 2008. Both *Total Primary Energy Supply* (TPES) and *Electricity consumption per capita* are given, in kW. (Source: International Energy Agency, www.iea.org)

	Population (millions)	TPES/capita (kW)	Electr/capita (kW)	GDP/capita (USD × 1,000)
USA	304.5	10.0	1.430	38.6
Netherlands	16.4	6.5	0.760	27.3
France	64.1	5.5	0.772	23.6
World	**6,687.9**	**2.4**	**0.287**	**6.1**
China	1,325.6	2.1	0.245	2.0
India	1,140.0	0.72	0.060	0.72
Bangladesh	160.0	0.23	0.023	0.46

The last column gives the Gross Domestic Product per capita, which may be used as an indication for the standard of living. Obviously, energy use is related to the standard of living. But there is no steadfast correspondence between the two, as is illustrated in the graph on p. 66. Countries may have similar GDPs but differ substantially in energy consumption. For example, Japan and the USA have about equally high GDPs per capita, but Japan has only half the energy use per person. In contrast, Saudi Arabia has a relatively low GDP but high energy use. Russia has a higher energy use per capita than the UK, Germany or France, while its GDP is only about one eighth of these countries.

From the graph, we can also conclude that a large group of industrialised countries such as the UK, Japan, Germany, France, the Netherlands and Russia have a higher energy use per capita than the world average by a factor of 2.5 to 3 (this is even a factor of 5 for the USA and Canada). The lesson here is that total world energy use would immediately become 2.5 to 3 times as large if every person in the world would have access to the same amount of energy as the people in most industrialised countries.

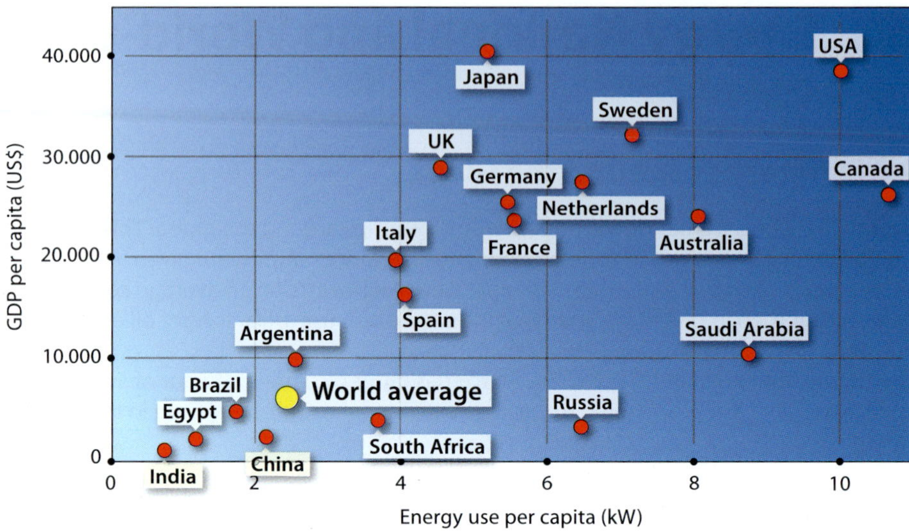

Gross domestic product per capita and energy use per capita for various countries. Many developing countries (not shown) are positioned in the lower left corner of the graph, around Egypt and India. Source: 2008 Key World Energy Statistics from the International Energy Agency.

B | Energy Use in an Industrialised Country: the Netherlands

By looking at a specific case in detail, much can be learned about energy use in industrialised countries. We will look at the Netherlands as an example.

The energy use per capita in the Netherlands is about 6.5 kW on a continuous basis. Expressed in terms of oil, this corresponds to 15 litres of oil per day. Note that this is about 60 times as much as the energy contained in our daily food, which was found to represent a quarter of a litre of oil. An immediate conclusion is that it is absurd to think that simply growing more food would make a substantial contribution to global energy use. So, 'solar fuels of the first generation' as they are called (fuel from corn, wheat, rape seed, sugar beets or sugar cane) can only modestly contribute to the current level of energy use in industrialised nations. We will return to solar fuels in chapter 8.

Energy supply

Where does the energy that is used come from? In 2008 in the Netherlands, the sources of energy were as shown in this diagram:

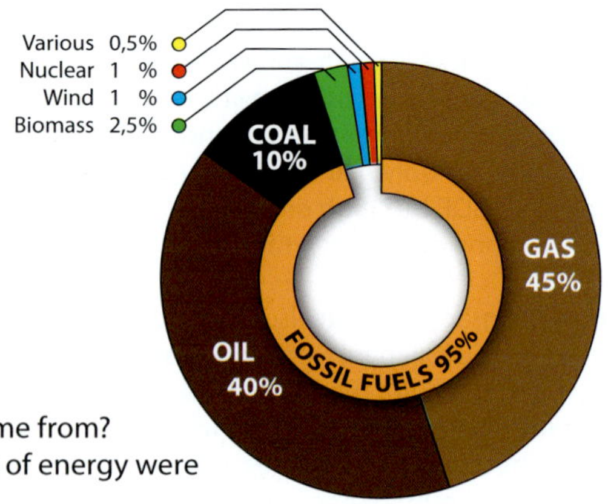

We notice that more than 90% of Dutch energy still comes from fossil fuels: natural gas, oil and coal. About one fifth of the natural gas and all of the coal go toward the generation of electricity (see below).

The energy balance of a country can be quite complicated. We should realise that there are imports and exports; there are reserves that increase or decrease; oil can be refined and then exported in a different form. Moreover, electricity can be generated by big power plants connected to the grid or by smaller units for local use, and so on. A breakdown of the energy use is therefore not always unambiguous.

Total energy use

Where and by whom is energy used? The answer to this question depends on what exactly we want to know. For example, in traditional thermal power plants, most of the energy is released as waste heat. Should we consider this heat as used by the plant or by the consumers? And what if this heat is being re-used to heat homes? Another example is the use of oil as a raw material for the chemical industry in the production of plastics, etc. Should we take this into account when we do our energy bookkeeping?

First, let us look at *where* the energy is used.
An approximate breakdown for 2008 is the following:

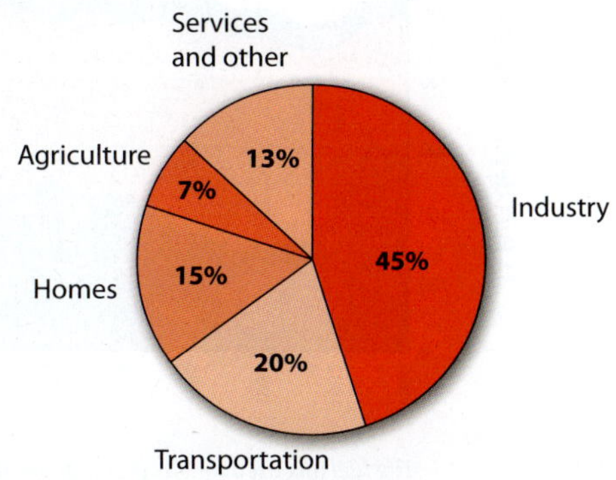

We may also wonder *for what purpose* energy is used. In the Netherlands, by far the largest part is used for heating homes and offices: about 45%. In many countries with a comparable climate this pattern is rather similar. So it is essential to consider home and office insulation as a prime strategy for increasing energy efficiency. This was discussed in great detail in chapter 4. Electricity generation represents one-quarter of the total energy use. Let us consider this in greater detail.

Electricity

The total electricity-generating capacity of the Netherlands in 2008 was 25,000 MW_e including co-generation (the subscript 'e' for electricity is used in this section to avoid possible confusion between the actual *electricity* produced and total energy used in the generation process).
The *average* electricity consumption was about 13,000 MW_e, slightly more than half the maximum generating capacity. This corresponds to the total power of 13 large power plants continuously operating to produce 1,000 MW_e each.

Average electricity consumption *per person* on a continuous basis was 760 W_e.
Note that, for domestic consumption, we calculated 400 W_e per household, or 180 W_e per person. This means that roughly one-quarter of the total electricity output is used in the home and three-quarters is used elsewhere.

Railroads consume 1.5% of total electricity. Streetlights along roads and lights in public spaces also represent an estimated 1.5% of the total electricity consumption.

Which of the primary energy sources are used for electricity generation? The fuel mix of primary energy can vary substantially from year to year because of price changes on the global energy market. The figure below represents the situation for the period 1980–2006. We see that the largest energy source is natural gas (red area) while coal (black) is also used extensively. Smaller contributions include heat from a garbage processing plant (small red area on top), biomass (orange), nuclear (blue), and industrial gas (green). Wind and solar do not yet appear on this graph; their joint contribution was still minute at this point, only 3% in 2006, but it has risen substantially in the following years.
The contribution of nuclear energy was about 4%. Nuclear energy as a source will be treated more extensively in chapter 11.

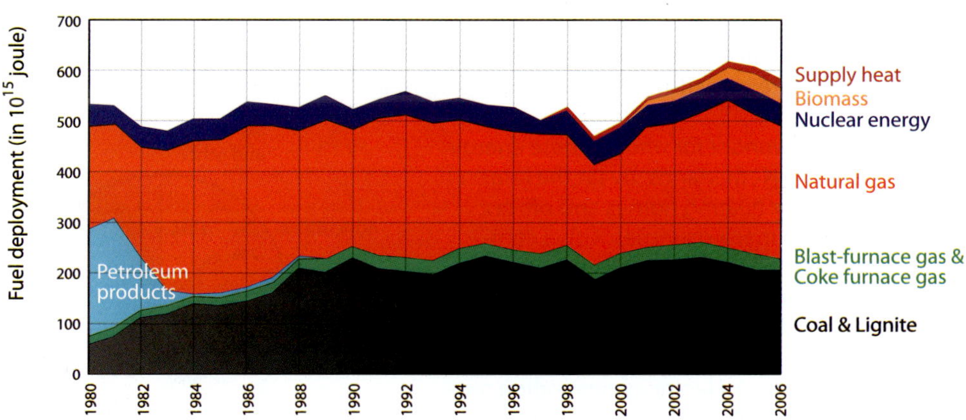

The fuel mix used for electricity generation in the Netherlands over the past few decades. We see that coal and natural gas are the foremost fuels. A substantial part of the natural gas is used in co-generation plants.

Co-generation

As discussed earlier, more than 50% of the energy used in thermal power plants is discarded as heat. This is because a fundamental law of physics limits the efficiency of the conversion process (see chapter 2). This heat is usually lost in cooling towers or in a river. However, this heat could be used for heating purposes, thus enhancing the efficiency of the entire process. Small, decentralised power plants are in a good position to utilise this process. They combine the generation of electricity with the generation of heat or, in other words, 'Co-generation' or 'Combined Heat and Power' (CHP). If heat is taken into account, the overall efficiency of co-generation could be nearly 100%. In the Netherlands, roughly half of the electricity is generated this way.

C | Europe

Total energy use in the EU-27 in 2008 was 73.4×10^{18} J (or 1,751 Mtoe; source: International Energy Agency). The breakdown into the various primary energy sources is shown in the pie chart below.

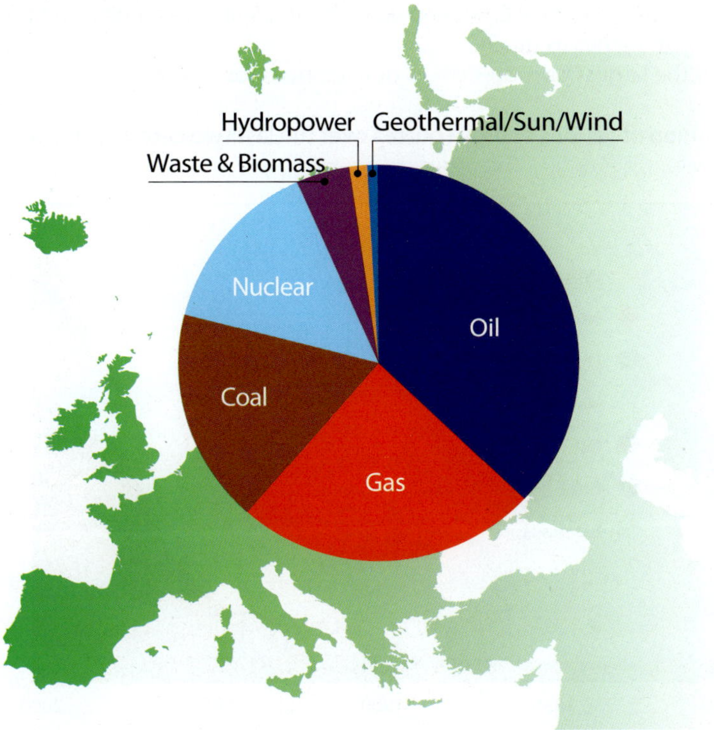

Contributions to the total primary energy supply in the EU in 2008 (Source: International Energy Agency: www.iea.org). The nuclear contribution includes the heat released.

The chart illustrates, for example, that:

- By far the largest portion (almost 80%) is still derived from fossil fuels.
- Nuclear energy contributes 14% (note that this includes the heat released).
- Renewable energy contributes 9%.

Note that nuclear energy and hydropower are used exclusively for electricity generation. If their contributions were to be expressed as a fraction of the *electricity* consumption, their relative contributions would be significantly larger. This is because electricity production in the EU comprises only about one-third of total energy use. If we look at *electricity* production, hydropower contributes 10.6 % and nuclear energy contributes 27.8%.
This is a frequent source of confusion, because people sometimes do not distinguish between energy and electricity. We will return to this more extensively when we discuss the world situation.

D | The World: Supply and Demand

Total energy use

World energy use in 2008 was 514 × 10^{18} J (Source: International Energy Agency: www.iea.org). Given a world population of 6.7 billion, this is 77 GJ per person per year. As seen in section B, a typical European uses about 2.5 times as much, and the average American about 4 times as much.
Again, by far the largest portion, 80%, is derived from fossil fuels.

The contribution of various energy sources over the last two centuries is shown in the graph below.

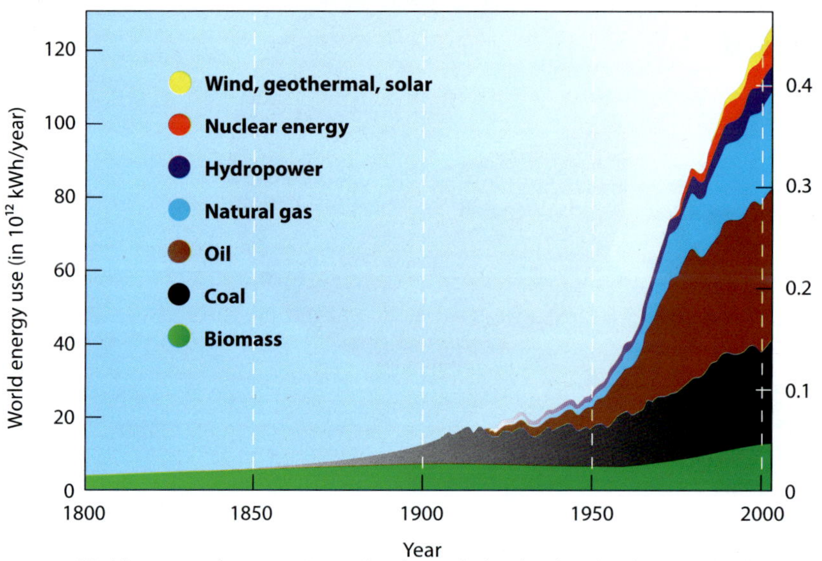

World energy use by source. Notice that during the last few decades about 80% has been derived from fossil fuels. Nuclear energy and hydropower contribute about equally.

Electricity

World *electricity* consumption in 2008 was 16.8 × 10^6 GWh. It was primarily generated by coal (41%), natural gas (21%), hydropower (16%), and nuclear power (14%) plants (source: IEA).
What portion of *total* energy use does this represent? This depends on how we look at it. If we simply convert the number of kilowatt-hours into joules, electricity is only 15% of total world energy usage. But that may be short-sighted. Most of the electricity is generated in thermal power plants (fossil fuels and uranium). If all electricity were generated from fossil fuels, electricity generation would consume about 35% of the total fossil fuel consumption. This is because the average conversion efficiency (from primary fuel to electrical outlet) is only about one-third. The rest, two-thirds, represents waste heat, which is sometimes re-used to heat homes ('Co-generation' or Combined Heat and Power), but in most cases it is discarded into rivers or lost in cooling towers.

This already makes things complicated. Moreover, some of the electricity comes from hydropower stations (16% in 2008). In this case there is no waste-heat problem. But if it became necessary to generate this amount of electricity using fossil fuels, we would have to spend about three times as much energy. This would be an argument in favour of multiplying the hydro-power-generated electricity by a factor of 3 if we were interested in assessing its contribution to the total world energy use.

The way we calculate energy use causes a great deal of confusion. One example is the nuclear energy debate. Nuclear and hydropower generate about the same amount of electricity each (14 and 16%, respectively, in 2008). So, if we assume that electricity represents about one-third of world primary energy use, both represent about 5% of *total* energy consumption in the world. But if we include the waste heat generated in nuclear power plants (even if it remains unused), nuclear energy would be three times more important than hydro power.
The International Energy Agency, for example, calculates its figures for electricity production in thermal power plants using the energy content of the electricity. Hydropower, wind, and other non-thermal forms of electricity production are accounted for similarly. However, for nuclear energy, the primary energy equivalent is calculated from the gross electricity generation by assuming a 33% conversion efficiency, *i.e.*, the energy content of the nuclear electricity is multiplied by a factor of three.
The message here is that one should handle these kinds of figures with great care (even those quoted in this book).

Fossil fuel reserves
How long will it take before fossil fuel supplies are totally depleted, assuming that world energy use were to remain at the present level AND that climate constraints will not prevent us from using the rest of the total supply?

The table on p. 72 presents the annual production of oil, natural gas and coal, calculated using the same unit, the joule, for easy comparison.
The second column includes the estimated 'proved reserves'.

Although there is no single, commonly accepted definition of proved reserves, a commonly used description is: 'those quantities that geological and engineering information indicates with reasonable certainty can be recovered in the future from known reservoirs under existing economic and operating conditions' (see www.bp.com).
Obviously, the 'proved' reserves are only a part of the 'probable' and even a smaller part of the 'possible' reserves. In turn, those are only part of the fossil fuels ultimately created by photosynthesis and subsequently turned into oil, coal and natural gas by a variety of complex fossilisation processes.

The table on p.72 indicates that oil and natural-gas reserves are comparable, but since oil is consumed more rapidly than gas, oil would be depleted first. Based on these numbers, both oil and gas would last for about half a century. In comparison, coal would last about two centuries because the reserves are larger than the oil and gas reserves combined.

It should be noted that large supplies of *shale gas* must probably be added to the reserves. This is gas that is tightly trapped in shale at depths of mostly 1.5 to 5 km, and consists mainly of methane. The technology of shale gas production is still in development. It usually involves hydraulic fracturing, or 'hydrofracking'. This technique uses small explosives to cause cracks in the shale. Subsequently, water mixed with sand is pumped into the cracks under high pressure. The gas then flows up the well, along with the fluid. Details can be found, for example, in *The Shale Gas Shock* by Matt Ridley (GWPF report 2, 2011).
Shale gas reserves may be found abundantly world-wide on all continents. It is not yet clear whether these reserves can economically be exploited on a large scale, and what the environmental drawbacks will be, but shale gas may cause a revolution in the gas industry.

Estimated 'proved reserves' R and yearly production P (in 10^{21} joules and 10^{21} joules/year, respectively) in 2009. The ratio R/P directly yields the number of years the supplies would last at constant consumption.
(Source: PB Statistical Review of World Energy, June 2010; www.bp.com)

	R (10^{21} J)	P (10^{21} J/y)	R/P (years)
Oil	7.6	0.166	46
Natural gas*	7.1	0.113	63
Coal	24.2	0.143	170
	---------	---------	
TOTAL	38.9	0.422	92 (= 38.9 / 0.422)

* This does not include shale gas (see text).

One should bear in mind that fossil fuels – notably oil and gas – also serve as essential raw materials for the chemical industry. It is not so easy to replace those with alternatives.

It is interesting to consider energy use in a geological perspective. Fossil fuels are basically formed as the product of photosynthesis, over a long period of time, probably a few hundred million years. However, if we do the calculation using the solar energy flux, the fraction of the globe covered by plant life and algae, and the efficiency of photosynthesis, we find that the total amount of fossil fuels represents a photosynthesis yield of roughly only a century. Given the above time span we must conclude that, during the era of fossil fuel formation, only a tiny fraction of the 'harvest' (on the order of a millionth) has fossilised.

During many centuries, the world's population has survived without ever using any fossil fuels – until around 1800, the start of the industrial revolution. It is remarkable that mankind is now in the process of consuming the total supply of fossil fuels, which needed a few hundred million years to be created, in just a few hundred years. Our consuming speed is a million to one, so to speak.

The historical perspective is schematically illustrated in the graph below. It shows the consumption of fossil fuels over the period A.D. 1000 to 2500. The data up until 2005 are based on facts. From there on an extrapolation has been made to 2050 where the level is assumed to be equal to *present* energy use for an average person in Western Europe, multiplied by the expected world population by 2050 (*i.e.*, 9 billion people). This would mean that the average per-capita consumption of fossil fuels in Western Europe would have to remain constant, and that the average person in the USA and Canada would have to reduce consumption by half, and that developing countries would increase to present Western-European levels. This does not sound like an unreasonable scenario.

The result would be an increase in per-capita use by a factor of 2.5 to 3 (see page 65) and an additional factor of 1.4 because of increases in world population. In total, this would lead to an increase in world energy use by a factor of at least 3.5 compared to 2005.

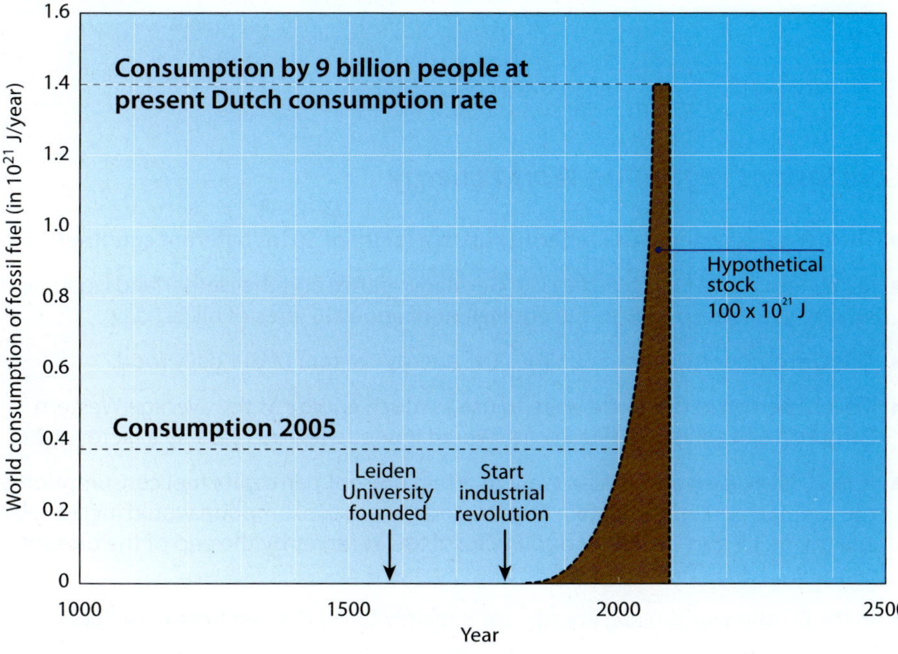

Super-simplified model illustrating the fossil-fuel era.

Upon reaching this level, let us assume that world energy use remains constant in the subsequent period. And let us be optimistic about total fossil-fuel reserves by assuming that they are larger than the table on p. 72 suggests by a factor of three (yielding 100×10^{21} J), and that no climate constraints prevent us from using them all. The resulting graph we would get is shown in the figure on p. 73.

We see that we would run out of fossil fuels late this century. Remember that we have been quite optimistic about present reserves.

Of course, this is by no means an informed prediction. The future will certainly be different. But it illustrates that we live in an extraordinary era: the fossil-fuel era, a very short period in the history of mankind. Note that civilization preceded the fossil-fuel era by several millennia. Even some European universities were founded centuries before fossil fuels were being used. The fossil-fuel era will last no longer than a few centuries. This is less than 0.1% of the time that human beings have thus far inhabited the globe.

The figure also shows that we should hope that not all of humanity will copy the energy-intensive lifestyle of the present developed world. If we look at this from an ethical point of view, we may want to question the extent of our present-day energy use.

Conclusions regarding World Energy

- The per-capita energy use differs by up to a factor of 50 for different countries.
- In Western European countries like Germany, France and the Netherlands, per-capita energy use corresponds to the equivalent of about 15 litres of oil per day.
- This corresponds to about 60 times the energy content of our daily food.
- If every person in the world were to use as much energy as the average Western European citizen, global energy use would immediately increase by a factor of 2.5.
- If, in 2050, every world citizen were to attain present per-capita fuel consumption of the average person in Western Europe, total world consumption would increase by a factor of 3.5 and we would deplete fossil fuel reserves by the end of the present century.
- In the EU and world-wide, energy use is mainly (80%) derived from fossil fuels.
- World fossil fuel reserves, created over a period of several *hundred million* years, may be totally consumed within a few *hundred* years.
- Oil and natural gas are indispensable as raw materials for the chemical industry.

7 | CO_2 and Global Warming

There is an intimate link between energy use and climate. This is due to the 'greenhouse effect' which everybody knows about. But not everybody realises that there was already a 'natural greenhouse effect' long before mankind started to spew large quantities of CO_2 back into the atmosphere. So let us first look at the 'natural greenhouse effect'.

A | Natural Greenhouse Effect

Planet earth is a kind of spaceship orbiting around the sun. Outer space is a vacuum, so that heat exchanges occur only via radiation. This means that the temperature of the globe is entirely determined by a radiation balance. The globe is heated by the sun (the only nearby star) and reaches an equilibrium temperature at which the total heat absorbed equals the total heat emitted.

On the one side we have incoming solar radiation. This is easily calculated by using a bit of high school physics (see box 'Radiation balance'). However, not all solar radiation intercepted by the earth is absorbed. Some portion of it is reflected and does not contribute to the energy balance. This part, determined by the reflection coefficient or 'albedo' (whiteness) of the earth, is rather well known, and has been verified by satellite data. The remaining part is absorbed, and represents the net incoming heat flux.

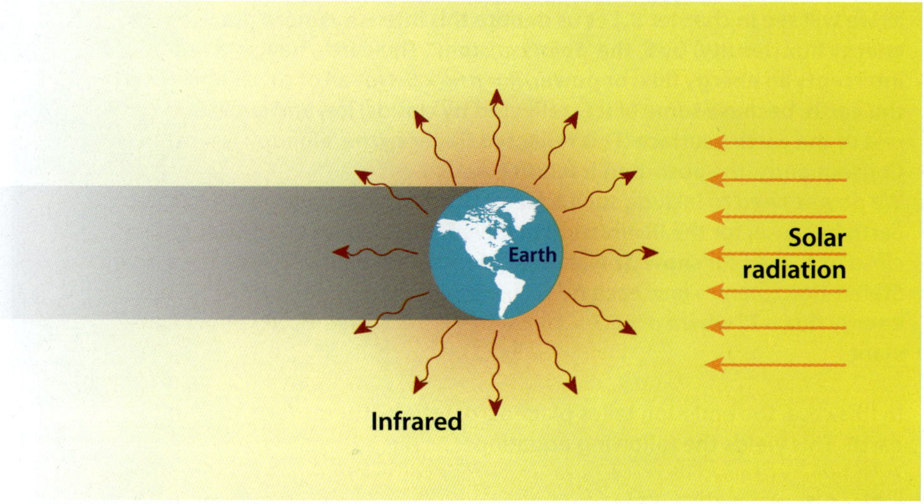

Schematic picture of the balance between incoming and outgoing radiation, which determines the earth's temperature

On the other side we have heat that is radiated by the earth back into space. This is 'far infrared' radiation, as explained in chapter 4 (see figure p. 35). It has a wavelength around 10 μm, about 20 times as long as visible light. At these wave lengths almost all surfaces are 'black', *i.e.*, they radiate heat with 100% efficiency, just like a black body. This makes the calculation process somewhat easier.

From such an elementary radiation balance, we find an average temperature for the earth of 255 K or –18°C (see box). This is substantially lower than the actual temperature of 288 K (or +15°C). The two differ by 33 degrees!

This difference in temperature is caused by the atmosphere. It is not so much the incoming radiation that is hindered by the atmosphere. It is the *outgoing* far infrared radiation that is largely absorbed. The absorption of the relevant IR radiation is mainly caused by a few molecules that have three or more atoms such as water (H_2O), carbon dioxide (CO_2), methane (CH_4) and a few others. They can absorb this radiation because their vibration and rotation energies are in the right energy range for absorbing far IR radiation. Simple diatomic molecules like nitrogen (N_2) are incapable of doing this. It is important to note that each absorbing molecule has its own specific absorption spectrum, *i.e.*, it absorbs specific parts of the spectrum. Examples of a few of these absorption spectra are shown in the figure on p. 77.

Radiation balance

The intensity of solar radiation when it enters the atmosphere is 1,350 W/m^2, as we will see in chapter 8. Let us denote this intensity (more precisely, the energy flux density) by *S*, the '*Solar constant*'. The earth, having a radius of *R*, intercepts an energy flux, or power, $P = \pi R^2 \times S$. Not all of this is absorbed by the earth, because some of it is reflected by clouds, ice, and snow and by the rest of the earth's surface. This reflected fraction (the 'albedo') is written as *r*. Consequently, the absorbed fraction equals $1 - r$.
We do not need to include such a factor for the radiation emitted by the earth because, for the IR emission involved, the emission coefficient is very close to 1, even for snow. The amount of radiation emitted is determined by Stefan-Boltzmann's law: each m^2 of a surface with a temperature *T* emits an energy flux σT^4 where $\sigma = 5.7 \times 10^{-8}$ W m^{-2}K^{-4} is Stefan-Boltzmann's constant.

In this case, the emission takes place across the entire surface $4\pi R^2$ of the earth. This yields the following equation:

$$\pi R^2 \times S(1 - r) = 4\pi R^2 \times \sigma T^4.$$

Using $r = 0.30$ we find $T = 255$ K or –18°C.

Radiation balance

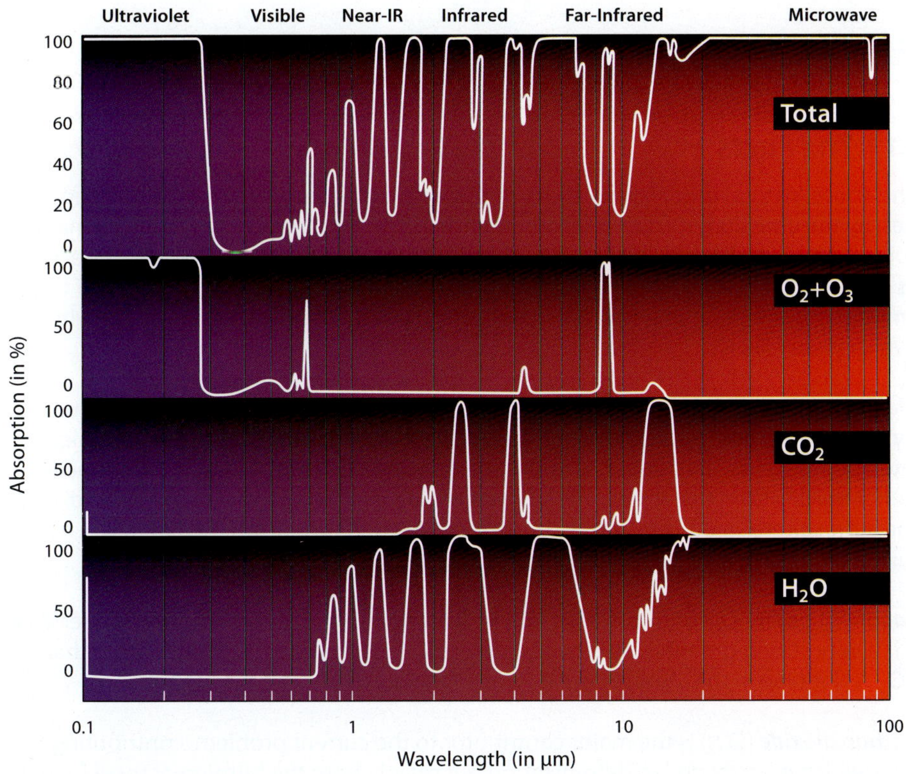

Absorption spectra of a few 'greenhouse gases'. Note that the absorption of these gases is mainly in the infrared region and that the peak of the spectrum emitted by the earth is at 10 micrometres, as shown in the figure on p. 35.

So, the situation is quite comparable to a greenhouse, where the sunlight enters almost unhindered, while the thermal radiation from inside is largely absorbed by the glass. In the case of the earth, some polyatomic gases take over the role of the glass. Hence the name 'Greenhouse effect'.

How does the surface of the earth heat up during this process? The gases that absorb the radiation also emit it, just like the glass panels of a greenhouse. This emission occurs in all directions, so that half of the radiation will go skyward, into space, and the other half will be directed downward and heat up the earth.

Of course, the real temperature balance of the earth's atmosphere is more complicated than the simplified model discussed here. The atmosphere is not a thin layer like a greenhouse's glass. Not only will the surface of the earth be heated but also the troposphere (the lower 10 km of the atmosphere). In contrast, the upper atmosphere will actually cool down owing to a net radiation loss.

The bottom line is that the 'natural greenhouse effect' makes the average temperature on earth 33 degrees warmer than it would otherwise be. "Fortunately!" one might add, because otherwise we would all be freezing.

B | Man-made Greenhouse Effect

B1 | The Facts

The above 'natural greenhouse effect' is currently being enhanced by human activities, by the emission of additional greenhouse gases into the atmosphere. The major source is carbon dioxide (CO_2), as detailed below, but other gases also play a role, such as methane (CH_4) and laughing gas (N_2O). Water (H_2O) should also be included if the amount of water vapour in the atmosphere is increased because of human activities (which is very likely).

Methane (CH_4) contributes about 15% to the greenhouse effect. It is primarily emitted by cattle (as flatulence), from manure and from leaks in natural gas pipelines. Calculated per molecule, it is about 20 times as effective as CO_2. Fortunately, it disappears rather quickly from the atmosphere, over a period of about 10 years, since it oxidises in the atmosphere. In other words, it 'burns' with oxygen producing the much less harmful CO_2 and H_2O.

Laughing gas (N_2O) contributes about 8%. Its primary sources are manure and fertilizer. It is a very strong greenhouse gas; per molecule, it contributes 300 times more than CO_2. Its lifetime in the atmosphere is about 100 years.

Carbon dioxide (CO_2) is the major contributor to the current problem, contributing 70%, which is partly due to deforestation but mainly from the burning of fossil fuels. In fact, the CO_2 is merely returning to its origins. Millions of years ago, before plants and algae, there was much more CO_2 (and less O_2) in the atmosphere than there is now. As plants began to grow, some of the CO_2 was bound by photosynthesis and converted into wood and other organic materials. Subsequently, over a very long period of time (lasting millions of years), this organic matter was eventually converted into fossil fuels. This process meant that most of the atmosphere's CO_2 literally went underground. This also means that the reverse, putting all CO_2 back into the atmosphere by burning all fossil fuel, will increase the CO_2 content in the atmosphere dramatically (see also the figure for carbon reservoirs on p. 80).

CO_2 is thus the major culprit when we talk of global warming. Its increasing concentration in the atmosphere has been meticulously measured since 1958 on Mauna Loa, Hawaii. The results are shown in the figure on page 79, where the last 50 years (showing the Mauna Loa results in detail) have been highlighted.
Various other stations around the globe (at the South Pole, in Australia, New Zealand, Alaska, etc.) yield essentially identical data, although over a shorter period of time.

A striking feature of the Mauna Loa data is that they also show the seasonal variations. In the spring a lot of CO_2 is absorbed by growing plants, whereas in the autumn much of it is released by the dying leaves and plants. (This raises the question of why the southern and northern hemispheres do not compensate for one another since they have opposite seasons.

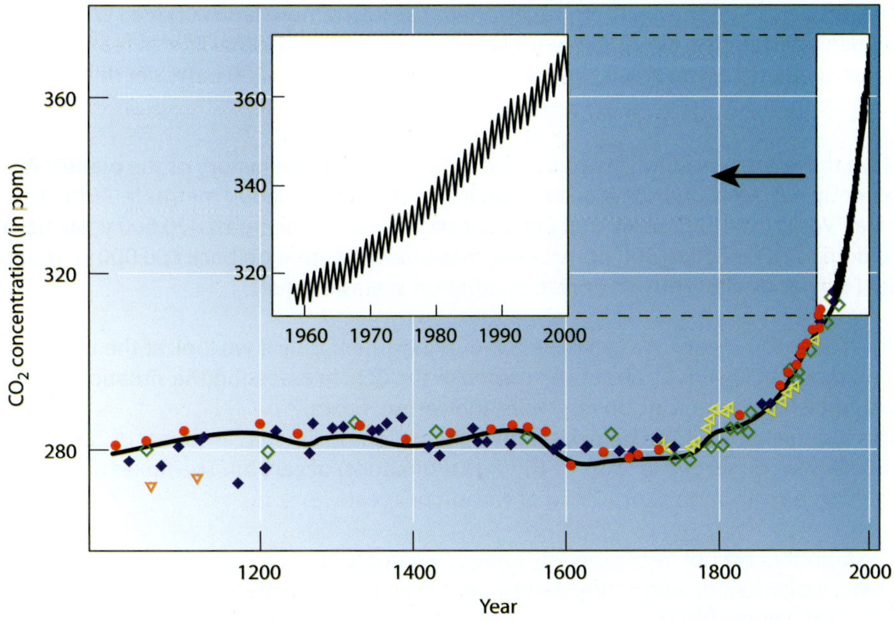

The concentration of carbon dioxide in the atmosphere as a function of time, in parts per million. The data from before 1958 were taken from measurements of air bubbles trapped in ice from various ice cores in Antarctica. Since 1958, direct measurements have been performed on Mauna Loa, Hawaii (inset). Note the seasonal variations of the data, which illustrate the experimental precision.

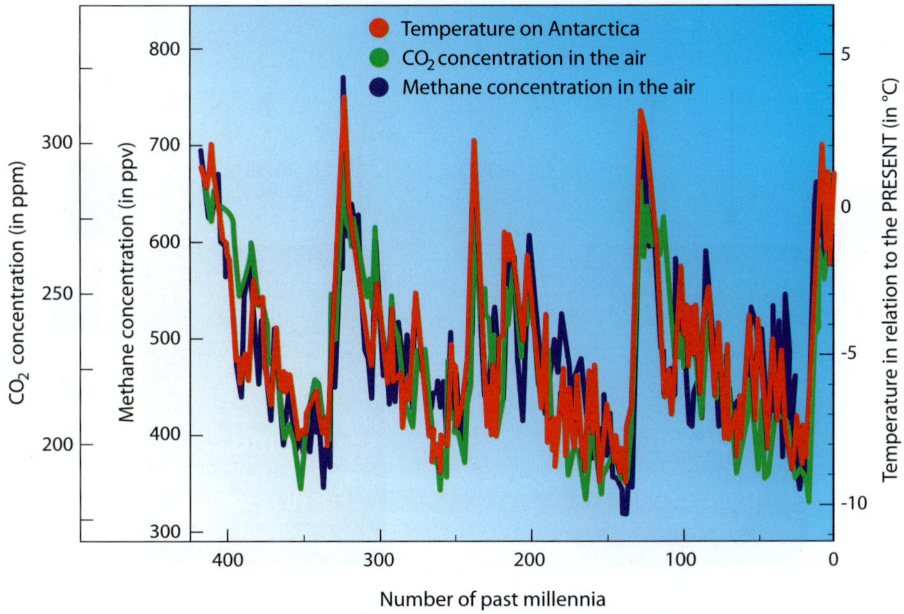

A glance back into the historical past: Temperature excursion from the present (red curve, scale on the right); CO_2 concentration (green curve, scale at far left) and CH_4 concentration (in parts per billion, blue curve) over the last 420,000 years. (Source: Intergovernmental Panel on Climate Change, IPCC)

The answer is that the northern hemisphere has much more land surface.) Without human interference, the graph would oscillate along a horizontal line. In reality, CO_2 content has increased drastically over the last century, from 280 parts per million (ppm) to more than 380 ppm.

Have these concentration levels ever occurred earlier in the history of the planet? Apparently not. Measurements taken from ice cores down to 3,600 metres in Antarctica (near 'Vostok station') show that CO_2 concentrations over the past 420,000 years have seldom ever exceeded 300 ppm. Other measurements that go back 650,000 years, present similar results. Thus, the present situation seems unique.

What does this mean? We can have a sound judgment only if we look at the numbers. If we discuss CO_2, it is all about the *carbon* in the CO_2. In assessing the situation we must distinguish between carbon *reservoirs* and carbon *fluxes*.
This is done in the figure below. It shows the various estimated amounts of carbon in the relevant reservoirs in units of 10^{12} kg (or gigatons) of carbon. The fluxes are indicated by arrows in units of 10^{12} kg of carbon per year.

The numbers are not very precisely known, except for the amount of C in the atmosphere, which can be accurately determined from its concentration and the total volume of the atmosphere.

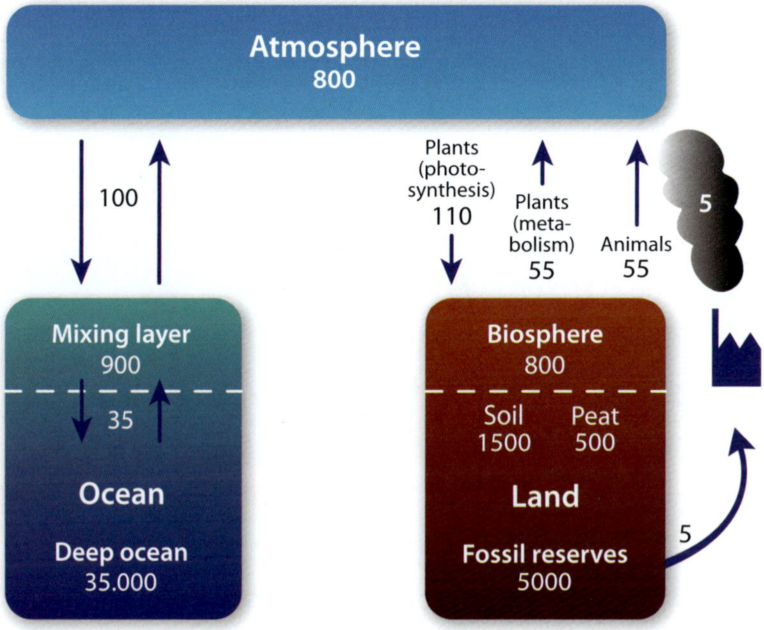

The various carbon reservoirs (rectangles, not to scale, in 10^{12} kg) and carbon fluxes (arrows, in 10^{12} kg per year) relevant to the issue of global warming.

When we look at the carbon *fluxes*, we notice that all of the major fluxes compensate for one another. While they may vary somewhat from year to year, they are approximately in balance when averaged over long periods of time. So the natural cycles have reached a long-term stationary state. This is true for both animals and human beings as well, because everything we eat represents some quantity of carbon that is removed from the air via photosynthesis, and is returned back to the atmosphere when we exhale.

However, what we notice is a small, new additional flux that disturbs the equilibrium. In short, the emission of CO_2 that occurs during the burning of fossil fuels exacerbates the greenhouse effect. The figure '5' is only an average. If we were to look solely at emissions, the figure would be somewhat larger (about 7), and when we look at the net atmospheric increase, the number is somewhat smaller (about 3.5). This is because roughly half of the emission disappears into the ocean and the terrestrial biosphere.

If we consider *reservoirs*, we notice that the atmosphere contains roughly as much carbon as the biosphere. So, to prevent the atmosphere's CO_2 levels from doubling, we have to double the area occupied by forests. Furthermore, we see that fossil fuel reserves (not *proved* reserves, but *geological* reserves) represent an amount of carbon that is much larger than the amount presently found in the atmosphere. If we were to burn all of this and assume that half of it would go into the atmosphere and the other half into the oceans, the half that goes into the atmosphere would still raise the CO_2 level by a factor of four.

The good news is that the ocean is a tremendous reservoir. It contains a great deal of CO_2, partly in the form of (bi)carbonates and partly as dissolved CO_2. This is comparable to carbonated drinks: when we open a carbonated-drink bottle we release the pressure and we see CO_2 bubbles emerging as a result. This process also works backwards: when the partial pressure of CO_2 above the oceans increases, the concentration in the water will also increase, since the two are in equilibrium. This creates opportunities for the possible storage of CO_2 in the oceans. This option will be discussed later (see Possible remedies, p. 85). Note, however, that by far the largest portion of carbon content in the oceans lies in the *deepest* areas where it only mixes very slowly with the relatively thin surface layer, as seen from the arrows indicating the exchange between the two.

B2 | Possible Consequences and Uncertainties

These are the facts thus far. The consequences are, to a certain extent, speculative, but considering the dramatic increase in CO_2 there will certainly be some kind of repercussion. Exactly what will happen is hard to tell. The climate is a very complicated three-dimensional system with many feed-back loops. But there is a reasonable degree of consensus among scientists regarding the results of the Intergovernmental Panel on Climate Change (IPCC), although there are questions about the precise details.

A few key results are shown in the graphs on the following pages.

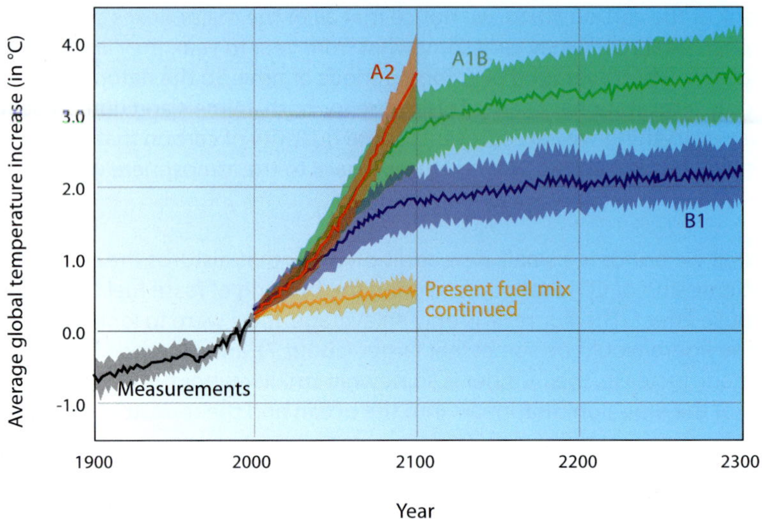

Average global temperature for different scenarios for CO_2 emissions, according to the IPCC. The upper graph (A2) assumes a future energy supply based mainly on fossil fuels. The blue curve (B1) assumes that renewable energy will become the primary energy source. Source: IPCC

Temperatures are expected to increase, averaged around the globe (see figure). However, there may be large geographic differences. These differences may affect the position of the various climate zones. This may mean that there will be less rainfall in the most fertile regions that are so important for agriculture, while there may be more rainfall in regions where the soil is unsuitable for agriculture. This could eventually jeopardise global food production.

As a result of increasing temperatures, sea levels would also rise, because of both melting of land ice and expansion of ocean water. The models predict a sea level rise of 20 to 50 cm by the year 2100, compared to 2000 levels. This is shown in the graph on page 83.

We see that the margins of error are relatively large, both in the temperature and in the sea level rise. This should not come as a surprise considering how complex the climate is with all of its feedback loops. The most well-known is the 'snow-ice-albedo': the reflection of solar radiation by snow and ice. If these ice and snow masses continue to melt, they will be replaced either by (darker) soil or by water. Both produce much less reflection than snow or ice. Indeed, while fresh snow has a reflection coefficient of almost 1, water has a much smaller reflection coefficient (only 2% at normal incidence). So, if snow and ice masses continue to melt, they will be replaced by a surface which absorbs solar radiation much more efficiently, producing ever-higher temperatures, which will accelerate the melting process of more ice and snow, etc. A real snowball effect, so to speak. Another positive feedback mechanism results from the role of water vapour as a greenhouse gas. There will be more water vapour because of higher temperatures, which will further exacerbate the greenhouse effect, which, in turn, produces more water vapour in the atmosphere, etc.

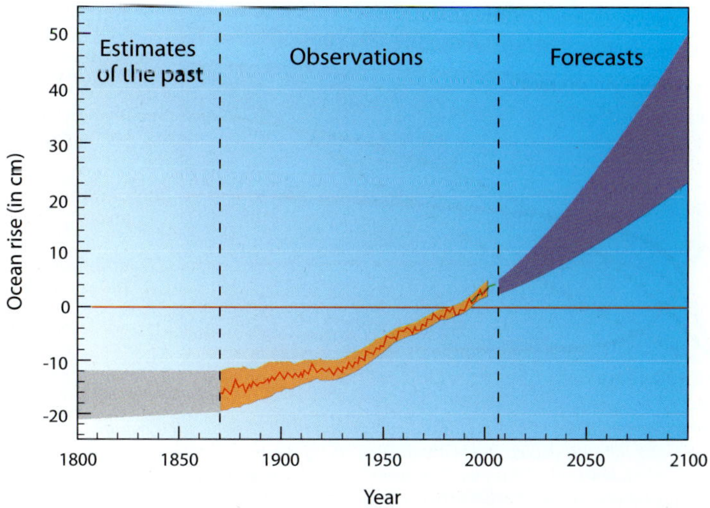

Average rise in global sea levels, recorded (1870–2008) and predicted (post-2008)
Source: IPCC

But there are also negative feedback mechanisms, or feedback systems not yet identified as being either positive or negative. For example: what happens with more clouds in the sky? On the one hand, they will reflect part of the incoming solar radiation, thus having a cooling effect. On the other hand, they will reflect more infrared radiation back to earth, which could lead to the opposite effect. It is not obvious which of these trends will prevail.
These feedback loops illustrate that the uncertainties remain substantial.

This uncertainty may seduce us into sitting back and waiting until we can be more certain. But this seems an unwise approach, if only because of the extremely large time scales involved (see p. 84). Even if we were to stop emitting greenhouse gases today (which is obviously impossible) it will still take a very long time before the atmosphere's CO_2 content returns to its pre-Industrial-Revolution state. We are talking about centuries here. So the problem is by no means analogous to the acid-rain problems in the 1970s. This problem was pretty much solved by stopping emissions (by requiring the installation of catalytic converters in all cars and using sulphur-free fossil fuels in the power plants), then wait for a rain shower and the problem was gone.

Exactly how much time is required for the various processes to reach equilibrium is illustrated by the figure on p. 84. We notice here that all of the effects are lagging behind CO_2 emission levels. This is especially the case for the rise in sea levels, which is caused by two effects: the expansion of water when temperatures increase (which itself lags behind CO_2 levels) and the melting of ice on land. Melting of ice in the air is a very slow process because melting requires a lot of heat and air is a poor thermal conductor. We may recall this when we observe piles of ice and snow at the end of winter. Even relatively small piles may survive far into the spring season.

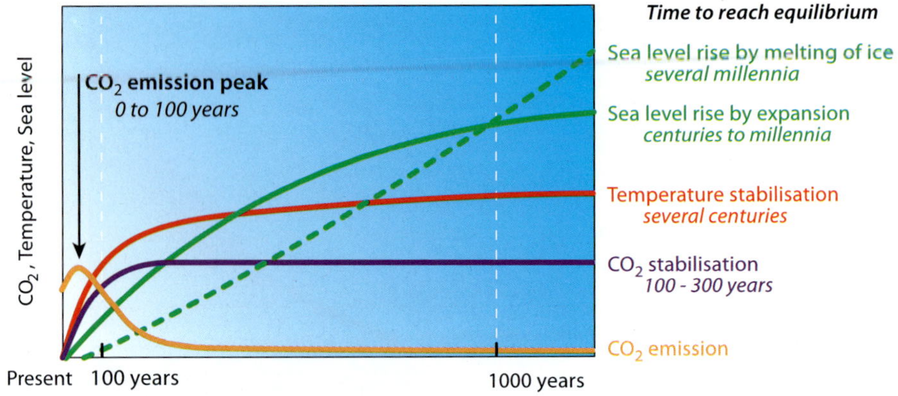

Time scales involved in global warming according to IPCC models. It is assumed here that CO_2 emissions will start decreasing in about 50 years from now (lower orange curve). However, it will take a few centuries for temperatures to stabilise (red curve) and perhaps even a millennium or more before sea levels will reach an equilibrium again.

Incidentally, floating ice does not lead to a rise of the sea level as it melts: Archimedes' law states that a chunk of ice has the same weight as the water it displaces, so after the ice melts it will still have the same volume as the water it was displacing before melting.

Why does it take so long for CO_2 equilibrium to be reached? Because in the ocean – the only natural reservoir that really matters – only the upper layer is involved over short periods of time. The deeper waters engage in a very slow exchange process with the upper layer (see the carbon reservoir picture on p. 80).

Many aspects of the global-warming issue contain uncertainties. For one thing, the climate system is extremely complex and not easy to model. Some features of this issue are not fully understood. An example is the precise role of aerosols, small liquid or solid particles that are suspended in the atmosphere. Even the historic temperature records may not always be reliable, since measurements may have been influenced over the years by expanding urbanisation or other human activities. The results of the IPCC models should therefore be used with caution, and uncertainties may be larger than suggested by the IPCC data.

The sun's variability is not a major source of uncertainty, as it normally causes temperature variations of only about 0.1°C, as will be pointed out in chapter 8 p. 91. However, deviations from the regular 11-year cycle of the sun's activity can cause larger fluctuations (see, *e.g.*, C. de Jager in Netherlands Journal of Geosciences, 87-3, 2008). Such deviations, with amplitudes on the order of 0.3°C over a 400 year-period, were probably the cause of the 'little ice age' in the late 17th century. Very few sunspots, which mirror the sun's activity, were observed at the time. Also, as of 2010 the number of sunspots had been decreasing, and the normal 11-year cycle seemed not to be continuing as expected. This could lead to an unexpected cooling effect, which may obscure part of the global warming in the period 2000–2008.

One should keep in mind, though, that such natural fluctuations occur on a timescale of tens of years, whereas the effects of increased atmospheric CO_2 occur on a timescale of centuries.

Volcanoes have only a temporary effect (1 to 2 years) on temperature. This effect is not due to CO_2 emission, which is negligible compared to anthropogenic emission. By far the most substantive effect is due to absorption of solar radiation by ash particles and various types of aerosols. This causes a heating effect in the upper atmosphere (the Stratosphere) by up to 1°C, as was recorded for example after the Pinatoba eruption in 1991. In contrast, this causes a temporary *cooling* effect in the lower atmosphere (the Troposphere) and on the earth's surface, since the solar radiation reaching the earth is attenuated by absorption in the stratosphere. This cooling effect is small, typically a few tenths of a degree, and vanishes after a few years.

B3 | Possible Remedies

Given the above details, it seems wise to limit the increase of atmospheric CO_2 as much as possible. The first remedy is to reduce emission by reducing fossil fuel use. Next, we can keep the emitted CO_2 from entering the atmosphere via carbon sequestration methods, *i.e.*, Carbon Capture and Storage (CCS).

Let's first look at the method for *capturing* CO_2. This is done most efficiently at large, concentrated sources, often called 'point sources'. The obvious targets are power plants, especially power plants fuelled by coal, which release relatively large quantities of CO_2 per unit of electricity produced (see p. 88). Also from an economic point of view, these cases are the most attractive.
Once captured, the CO_2 gas can be compressed and transported via pipelines. This is feasible up to distances of about 1,000 km.

The next question is: where do we store it? We have to find carbon reservoirs outside of the atmosphere. This is not easy, given the enormous quantities of CO_2 gas emitted.

Planting trees, for example, sounds nice, but is not sufficient. We have already noticed that the complete biosphere, including all of the present trees on earth, contains roughly the same amount of carbon as the atmosphere. So, if we want to compensate the rise of CO_2 in the atmosphere by a certain fraction, the amount of carbon stored in plants and trees has to increase by roughly the same fraction. This will not be easy to accomplish, given prevailing deforestation trends. Moreover, we need to realise that the ecosystem is in principle a closed cycle. Exactly the same amount of CO_2 that is currently being captured by photosynthesis will be released later as plants die. The natural ecosystem is therefore CO_2-neutral, unless we withdraw biomass from the ecosystem and build a new carbon reservoir. This could be done, for example, by using trees to make furniture, or timber piles to build houses on, as was done in cities like Amsterdam in the past. Using biomass as a source of energy rather than using fossil fuels also helps. If we ensure the replacement of the biomass withdrawn with new plants and trees, the net effect could indeed mean less CO_2 emission into the atmosphere.

Other options beyond the natural cycle are:
- Geological storage: placing CO_2 in underground reservoirs. The possibilities, both technically and economically speaking, remain unclear. Combining storage with oil and gas production seems an economically attractive option. The CO_2 is then injected into the wells and used to drive out oil or gas. This technique is called carbon sequestration by enhanced oil recovery (EOR) or enhanced gas recovery (EGR).

- Storage in the ocean. This seems an attractive option, given the fact that the oceans represent a very large carbon reservoir. They contain about 50 times as much carbon as the atmosphere (see figure p. 80), so their capacity is huge. Moreover, most of the excess atmospheric carbon dioxide will eventually be taken up by the oceans anyway. Note, however, that the active uppermost layer is relatively thin. We must therefore also address the deeper parts of the ocean.

Roughly speaking, there are two possible approaches to storing CO_2 in deep water. One is to inject the CO_2 into relatively shallow water, say, 1 km. This can be done directly at the source, such as a power plant. The CO_2 gas will initially rise as bubbles and then gradually dissolve in the water. This is sketched out on the left in the figure below.

The second option is to release the gases in deeper water. Here we are lucky because at a depth of 3 km the pressure is 300 bars. At such pressure levels, the CO_2 is heavier than water, so it will actually sink, become fluid, and gradually dissolve in the water. If time permits and the bottom is not too deep, it may even form a CO_2 lake (see the figure). Dumping CO_2 at such depths could also be done from a platform, not only from ships.

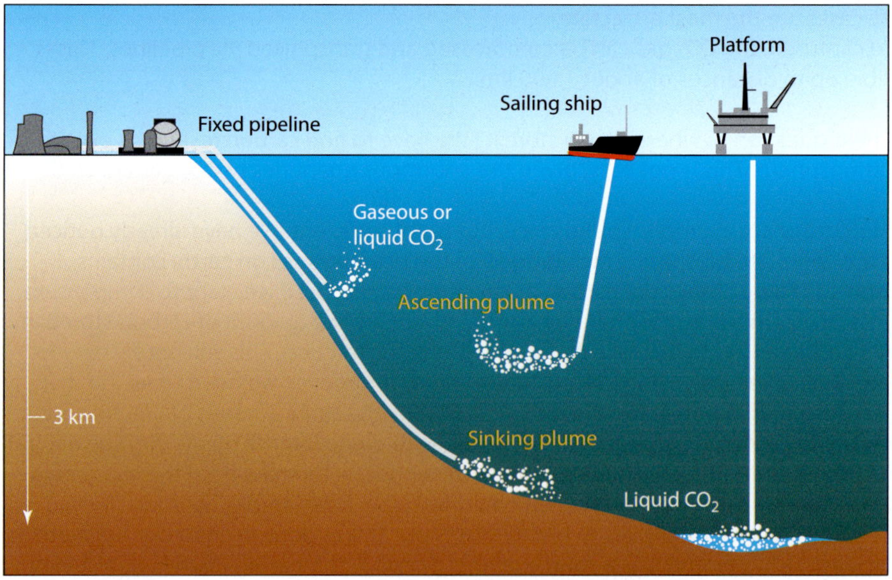

CO_2 storage in the ocean presented schematically. At deeper than 3 km, CO_2 becomes heavier than water and sinks (Source: IPCC Special Report: Carbon Dioxide Capture and Storage; www.ipcc.ch).

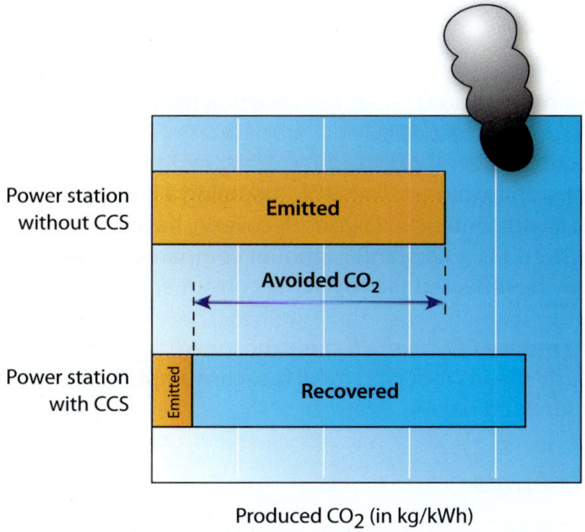

CO_2 emissions, in kg of CO_2 per kWh of electricity produced, for a power plant without (upper) and with (lower) Carbon Capture and Storage. The net emission reduction is the difference between the brown areas (source IPCC Special Report: Carbon Dioxide Capture and Storage, www.ipcc.ch).

Ocean currents offer an unexpected advantage. In Northern regions the relatively cold Atlantic waters sink, flow slowly southward in the ocean depths to resurface again between 100 to 1000 years later. They then begin their journey back to the North. This natural transport mechanism may offer possibilities for an elegant CO_2 transport system via the ocean's depths.

There are obviously drawbacks, for example, the change in acidity (the 'pH') of the ocean water. This will have an effect on the ecosystem and could lead to the extinction of various forms of marine life. The precise consequences of dumping CO_2 in the ocean are as yet unknown.

- The use of minerals. We can employ the chemical reaction of CO_2 with metal oxides that are found in natural silicate minerals. One example is olivine, a very common mineral, the chemical formula of which is Mg_2SiO_4 (or Mg replaced by Fe). The relevant reaction is

$$Mg_2SiO_4 + 2CO_2 \rightarrow SiO_2 + 2MgCO_3.$$

A drawback is that this reaction proceeds slowly. It is not yet clear whether this method is feasible on a large scale.

- Re-use in the chemical industry. This is, of course, the most elegant method, but totally insufficient if we look at the huge quantities involved.

B4 | Economic Aspects

All of the measures calculated to capture and store CO_2 that have been mentioned also require a certain amount of energy. Obviously, the net effect of the method used must be to reduce CO_2 emissions. So, what is the situation at power plants equipped with CCS?

According to IPPC data, a plant connected to a geological or ocean CO_2 storage system can reduce net emissions by 80 to 90%, with the precise percentage depending on the type of fuel used. This is schematically illustrated in the figure above. The bars represent *total* emission, which is larger in a CCS-equipped plant.

However, since most of the emission is captured (blue), net emission (brown) is much less for a CCS-equipped plant. The difference is 85%.

The next question is how much all of this will cost. The IPCC, assuming a geological storage system (possibly combined with Enhanced Oil/Gas Recovery), has calculated between € 0.02/kWh to € 0.04/kWh. This is a substantial amount compared to electricity *production* costs, but would only be reflected slightly in the consumer prices.

By far the most economical way to reduce CO_2 levels in the atmosphere is to reduce the emissions themselves. This obviously means reducing fossil fuel consumption. Exactly how the two are related will be addressed below.

B5 | Emission for Various Fuels

Fossil fuels vary in the amount of CO_2 emissions they produce during combustion. It is the 'C' in the hydrocarbons that produces carbon dioxide: each single C atom combines with oxygen in the air to produce one CO_2 molecule. Therefore, we can expect that natural gas, mainly composed of CH_4, will produce less CO_2 than the longer hydrocarbon chains C_nH_{2n+2}, like oil or petrol, for which n is around 10. Even worse, however, is coal, which is almost entirely made up of carbon.

These data are important when attempting to minimise CO_2 emissions for a certain level of electricity production. The table below shows that coal is the worst as far as emissions are concerned, while natural gas (mostly CH_4) is the best.

The number of kg of CO_2 emitted during combustion, per gigajoule of energy produced

Coal	89	kg CO_2 / GJ
Crude Oil	71	kg CO_2 / GJ
Natural gas	51	kg CO_2 / GJ

Sometimes it is useful to know the amount of CO_2 emitted *per unit weight of fuel* consumed. This can be easily estimated using basic high school physics (see box on the facing page). The most common cases are presented in the table below.

The amount of CO_2 produced during combustion of various fuels
(see also box)

1 kg of oil or petrol produces about	3	kg of CO_2
1 litre of petrol produces about	2.4	kg of CO_2
1 m^3 of natural gas (1 bar, 0°C) produces about	1.8	kg of CO_2
1 kg of coal produces about	2.6	kg of CO_2

Fuel and Emission

If we use some high-school science, it is really easy to estimate how much CO_2 is actually released during combustion. Here is how:

Petrol and oil

These are mainly hydrocarbons of the type C_nH_{2n+2} (alkanes). Since n is on the order of 10, we may represent 2n+2 simply by 2n, especially if we are interested in mass (remember that an H atom is light compared to a C atom: the ratio is 1 to 12). So we make believe we are burning C_nH_{2n}. This can be written as:

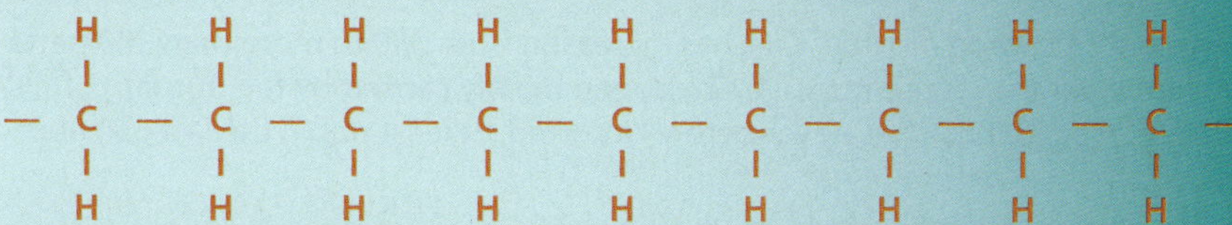

If we are interested in the mass of the CO_2 produced, we just look at the above scheme. Each vertical row (essentially CH_2, having a mass of 14 in elementary mass units) will form one CO_2 (with a mass of 44). Thus the mass of the CO_2 after combustion will be 44/14 times greater than the mass of the initial fuel. The number 44/14 or 22/7 may ring a bell. It is, quite coincidentally, the traditional approximation for π. So this estimate allows us to conclude that each kg of oil or petrol will produce 3.14 kg of CO_2, or, roughly speaking, 3 kg. If we want the result for a *litre* of fuel, we note that 1 litre of diesel is about 0.85 kg and one litre of petrol is about 0.75 kg. This alters the figures somewhat (see the table).

Natural gas

The calculation for natural gas is relatively easy if we use pure CH_4. Each CH_4 molecule will produce one molecule of CO_2. This is also true for a cubic metre, since equal volumes of gas at equal pressure and temperature have equal numbers of molecules. Now, 1 m^3 of CH_4 at standard temperature and pressure contains a number of moles given by 1,000 / 22.4 = 44.6 moles. Each of these has a mass of 44 grams. This yields 44.6 × 44 grams = 1.96 kg of CO_2 per m^3 of CH_4.
But natural gas is not pure CH_4. Standard natural gas contains 14% nitrogen, for example, and this does not produce any CO_2 since it contains no carbon. This lowers the emission slightly to 1.78 kg CO_2 per m^3 of standard natural gas.

Coal

If coal were pure carbon, each atom would combine with two O atoms to form CO_2. This would yield a total mass of 44 in elementary mass units, while the initial carbon had only 12. This would produce 44/12 or 3.67 kg CO_2 per kg of coal. But coal is not pure carbon, just think of the ashes that remain after combustion. This lowers the amount of CO_2 produced per kg. Standard coal turns out to produce 2.6 kg of CO_2 per kg of coal.

Conclusions regarding CO_2 and Global Warming

- The amount of carbon dioxide, CO_2, has increased from 280 to more than 380 parts per million over the last century. This has been mainly caused by the burning of fossil fuels. A similar rise has never before occurred, at least not in the last 650,000 years.
- Roughly half of the CO_2 emitted is absorbed by the ecosystem, while the other half remains in the atmosphere for centuries.
- Since CO_2 is one of the polyatomic gases that absorb infrared radiation emitted by the earth, it may have a substantial influence on the world's climate. The precise consequences are very difficult to predict. Uneven warming effects and shifting climate zones seem likely scenarios.
- According to predictions based on models produced by the Intergovernmental Panel on Climate Change (www.IPCC.ch) the *average* global temperature may rise by 2 to 4 degrees during this century. This may vary from region to region.
- Sea levels may rise by anywhere from 20 to 40 cm during the coming century.
- The sun's variability causes global temperature variations of only about 0.1°C.
- Volcano eruptions have a temporary (1 to 2 years) cooling effect. This is mainly due to absorption of solar radiation by ash particles and aerosols in the upper atmosphere.
- The ocean is by far the largest natural carbon reservoir. If CO_2 was dumped at a depth of more than 3 km, it would sink automatically and form temporary 'lakes'. A drawback is that ocean water is acidified by the uptake of CO_2, with as yet unknown consequences for marine life.
- Storing CO_2 on land is another option. This may be economically feasible if it can be combined with oil and/or gas production ('Enhanced Oil/Gas Recovery').
- According to IPCC calculations, the capture and storage of CO_2 in the vicinity of power plants will raise the electricity price by € 0.02/kWh to 0.04/kWh.
- When burning fossil fuel to produce heat, as is done in power plants, coal produces 89 kg of CO_2 per GJ, and natural gas only 51 kg per GJ.
- Burning 1 m^3 of natural gas produces about 2 kg of CO_2, burning 1 litre of petrol about 2.4 kg.

8 | Solar Energy

A | The Potential of Solar Energy

The sun is our energy source *par excellence*. Not only does the sun provide the energy that makes life on earth possible, it has also provided us with a nice piggy bank in the form of oil, natural gas and coal reserves.

How much energy *is* the sun radiating? This energy flux is truly unimaginable: 3.8×10^{26} watts. For people who would like to check this on the back of an envelope, the derivation can be found in the box 'How much energy does the sun radiate?'.

How immense this is, can be seen if we express it in quantities related to the human scale. For example:
- If the sun's energy flux were expressed in terms of barrels of oil and was nicely distributed among all human beings on earth, each of us would receive ten million barrels of oil *per second*. (We would all be rich in a second!)
- If it is expressed in terms of loss of mass, using Einstein's famous formula $E = mc^2$, we see that the sun loses four billion kilograms *every second*. This is more than half a kilogram for every person on earth. This may sound like an awful lot, but it is just a tiny fraction of the sun's colossal mass. In fact, given the solar mass of 2×10^{30} kg, it loses only 0.01% in a billion (10^9) years. The sun has been shining for about 5 billion years, and has about the same amount of time left. So the loss of mass is nothing for us to worry about.

A1 | Solar Energy on Earth

What does this mean for the planet earth? Given the distance to the sun, the solar irradiance is found to be about 1,350 watts per m^2 (see box 'How much energy does the sun radiate?'). Incidentally, this number remains fairly constant. Variations owing to fluctuations in solar activity are normally on the order of 0.1%. A bit more precisely: between 1978 and 2010 the solar irradiance has oscillated between 1,365 and 1,367 W/m^2 (source: R. van Dorland, KNMI, 2009). This does not noticeably affect the global temperature. The physics behind this is that the radiation flux is proportional to the fourth power of the temperature. The temperature will only vary by a quarter of the above 0.1 %, which corresponds to roughly 0.1 degree.

Our planet intercepts an amount of energy equal to the area of the 'global disc' times the above 1,350 W/m^2. If we do the calculations, we come up with 1.75×10^{17} watts, of which about 40% is reflected, primarily by clouds. The remaining 60% is still large: it corresponds to about 10,000 times the world energy use at present.
This leads us to a few interesting conclusions.

***Conclusion 1**: The amount of heat produced by human energy use is negligible compared to the solar influx. Therefore, heat produced by human activity does not have a directly noticeable influence on global temperatures.*

***Conclusion 2**: We only need to tap into a very small fraction of the solar energy to provide all the energy used by the world population at present.*

How much energy does the sun radiate?

We can easily calculate the energy flux emitted by the sun using a bit of high-school physics. The only data we need – in addition to the radiation law – are shown in the figure: the temperature T of the 'outer layer' of the sun and the radius R_s of the sun. The 'outer layer' of the sun, which emits the sun's radiation, effectively behaves as a perfect emitter (like a black body). The emitted energy is easily found if we apply Stefan-Boltzmann's radiation law, which states that the energy flux emitted per square metre by a perfect emitter is represented by σT^4, where σ is Stefan-Boltzmann's constant 5.7×10^{-8} $Wm^{-2}K^{-4}$. If we multiply this by the surface area of the sun, we will arrive at the total energy flux, or power P, emitted by the sun:

$$P = (4\pi R_s^2)\, \sigma T^4.$$

When we put the numbers in, we come up with 3.8×10^{26} W, which represents the total energy flux emitted by the sun.

If we want to know the level of solar radiation on earth, we need the distance d between sun and earth. This distance, by the way, varies over the year since the orbit of the earth around the sun is not a perfect circle. Around July 1 the sun is 3% *further away* than it is at around January 1. The figure shows the average distance.

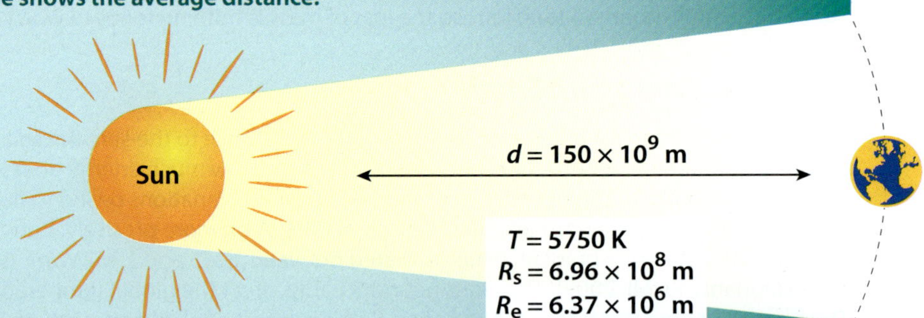

At distance d the total flux P found above is 'spread out' over a sphere with radius d, and a surface area A of $4\pi d^2$. So, per unit area, this yields (the 4π drops out nicely):

$$P/A = \sigma T^4 (R_s/d)^2.$$

When we put the numbers in, we come up with 1,350 W/m^2 for the amount of solar irradiance entering the earth's atmosphere.

If we want the total energy flux intercepted by the earth, we simply multiply this irradiance by the area that leaves a shadow behind the earth, or πR_e^2.

The above 1.35 kW/m^2 must be slightly corrected for the absorption of solar radiation by the atmosphere's gases. The amount of absorption, and, thus, also the solar energy flux actually reaching the earth's surface, depends on the angle at which the solar rays reach us. At normal incidence (*i.e.*, midday near the equator) about 20% is absorbed. The remaining 1 kW/m^2 reaches the ground, if there are no clouds. At higher latitude, say in Western Europe or the Americas, this amount is only slightly less. So we arrive at a simple rule of thumb: the sun delivers 1 kW/m^2 on a cloudless day if the sun is high. If the sun is somewhat lower, for example at 60° above the horizon, the distance the sun's rays need to travel through the atmosphere barely increases (to be precise, it increases as 1 / sin h, where h is the height of the sun above the horizon in degrees). So, in most countries, a maximum value of 1 kW/m^2 measured perpendicularly to the sun's rays, is a good approximation. This leads us to:

Conclusion 3: *On a cloudless day, when the sun is high, each m^2 receives 1 kW. This is the equivalent of a typical electric heater.*

Incidentally, we can estimate this value of 1 kW/m^2 fairly accurately by putting our hand close to a 100W light bulb, such that it feels like you are holding your hand out in the sun. Now, we assume that the light bulb radiates equally in all directions such that at distance d its power is distributed evenly over a sphere having radius d. We should come up with something of the order of 1 kW/m^2.

A2 | Yearly Yield of Solar Energy

In countries at a latitude similar to that of the UK, the Netherlands and Germany, the sun is low during the winter months. Even at midday, the path length of the solar rays through the atmosphere is much longer than in the summer months. As an example, at a latitude of 50° north, the path length of the sun's rays at midday is almost four times as long in the winter as in the summer. This means that the solar energy flux is appreciably attenuated. Its strength is a bit less than half the summer value, measured perpendicular to the sun's rays. Moreover, the days are much shorter in wintertime, which means the solar energy yield per day is further reduced.

For many applications, we are interested in the day-and-night average of the solar energy flux on a *horizontal* surface. More specifically, we want to know the *net* energy flux, which takes the attenuation by clouds into account. Obviously, this varies strongly over the seasons. As an example, in the Netherlands the net energy flux per square metre varies from 20 W/m^2 in the winter to 200 W/m^2 in the summer. The average is 110 W/m^2, which is also nicely seen in the figure on page 94. (Extensive data, including data for solar panels positioned at optimal angles, can be found on the website of the European Commission's PhotoVoltaic Geographical Information Service (PVGIS): http://re.jrc.ec.europa.eu/pvgis/solres/solreseurope.htm).

If we multiply the above 110 W/m^2 by the number of seconds per year (which is, quite coincidentally, almost exactly π times 10^7 seconds) we arrive at a yearly solar energy of 3.5×10^9 joules per square metre. This is the equivalent of about 90 litres of oil per m^2, or a 90-mm layer.

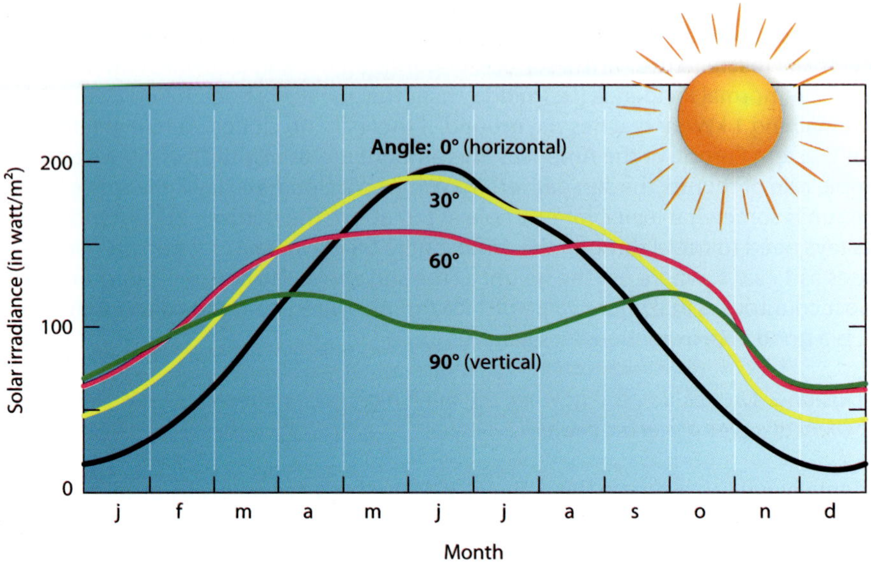

Average day-and-night solar energy flux per square metre on a surface facing South at various angles with respect to the horizontal (clouds have already been taken into account). These graphs refer to the Netherlands. They show that, on a horizontal surface, the energy flux in winter is very small. A normal window captures the most sunlight during the spring and autumn. The average over the year is 110 W/m^2.

This leads to the following conclusion:

***Conclusion 4**: If we collect all of the solar energy delivered to the soil in northwestern Europe for an entire year and express it in oil, by December 31 we would be up to our ankles in oil.*

If we consider solar panels or windows, we are interested in the amount of solar energy delivered to a surface *at an angle* throughout the year. For countries that are far from the equator, seasonal variations are large. As an example, this is shown in the figure above for the Netherlands.

It is seen that solar panels placed at an angle perform better than horizontal panels if we take an average over an entire year, although the differences are not that significant. However, it is often advantageous to achieve a large solar energy harvest during the winter (just think about heating, for instance). In this case, it is clearly preferable to have panels at an angle of, for example, 60°. Note also from the green curve that ordinary windows do rather well, especially in spring and autumn, when they deliver circa 110 W/m^2 as a daily average. Even in the middle of the winter they do pretty well. The reason is that, although the sun is weak, it shines almost perpendicular to the window. In the summer, the sun is so high that it hardly shines into our rooms. Moreover, it rises and sets a bit toward the north, and misses the south-facing window altogether in the early morning and late evening.

We have to realise that solar energy is also delivered when it is cloudy. After all, all light that continues to shine through the clouds represents energy. You can still take photographs, for example, although the exposure time is four times longer. This means that we can estimate the irradiance to be about 25% of the yield during a clear day.

The radiation is *diffuse*, however, which means that it cannot be concentrated by using mirrors. But we can still use the energy for solar panels and photosynthesis. In countries like the Netherlands, diffuse radiation contributes about 65% of total energy. An advantage of diffuse radiation is that it offers a lot of flexibility to architects and urban planners who want to employ solar energy.

Thanks to the contribution of diffuse radiation, the total year-round solar energy flux in countries like the UK, Germany and the Netherlands is between 100 and 150 W/m^2, which is not as bad as one would expect. In fact, it is only about a factor of 2 lower than most 'sunny' spots around the world.

This is illustrated in the figure below.

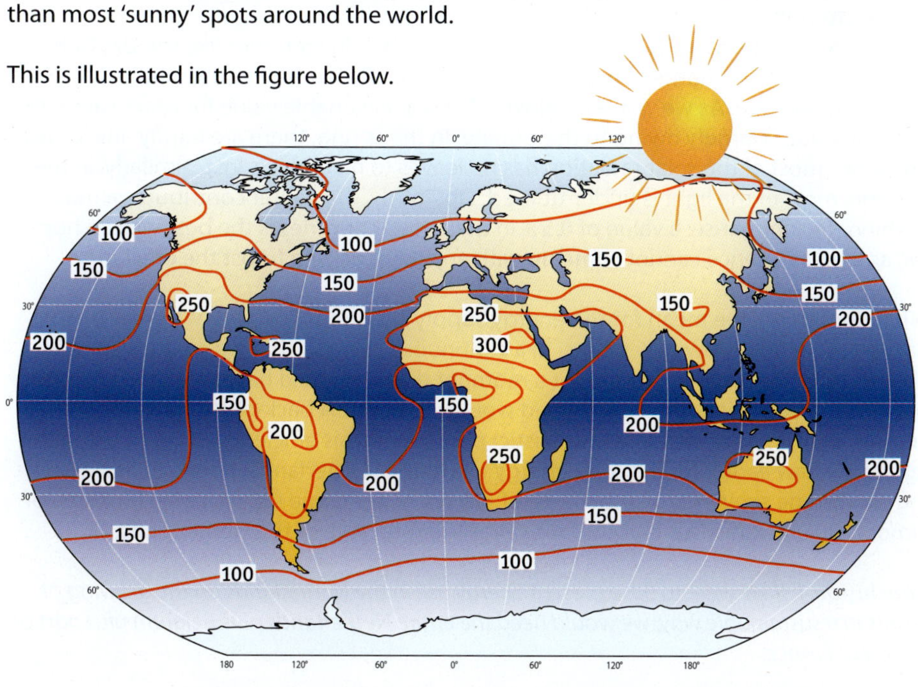

Average day-and-night solar irradiance over an entire year for various places around the world. The global average is 160 W/m^2. Countries like the UK and Germany receive 100 to 150 W/m^2, whereas the global maximum is 300 W/m^2. These numbers may be translated into the number of kWh per year by using: 1 W = 8.76 kWh/year.

B | Capturing Solar Energy by Various Mechanisms

Of the various possibilities for harvesting solar energy, let us first address the mechanisms presented by nature: harvest in a literal sense.

B1 | Trees, Plants, Solar fuels, Algae, Bacteria

Nature has developed a nice system. Using water and atmospheric carbon dioxide as ingredients, chlorophyll converts part of the incoming radiation into organic material, by means of photosynthesis.

The advantage is that the system itself handles storage. We can use the harvest at any time, in contrast to the 'harvest' of solar panels or photovoltaic cells. Products of photosynthesis are presently the only option we have for alternative fuels for transportation.

A drawback is that the efficiency is low. Under optimum conditions, the theoretical efficiency is 3% (see box 'Efficiency of photosynthesis'). Indeed, this value is only actually achieved in practice in a favourable climate, by plants such as sugar cane. In a moderate climate, the efficiency is much lower. Let us first look at this case.

Moderate climate

What is the situation like in a moderate climate such as in countries like the UK, Germany, and the Netherlands? The upper limit for the efficiency of photosynthesis is 1% as a yearly average. As we will see below, 0.5% is a reasonable value for practical cases. This low value is largely owing to the climate. In the spring, there are hardly any leaves on plants: most of the solar radiation has no leaves to hit and it is lost. Similarly, in the autumn, most of the light is under-utilised. Obviously, the winter contributes almost nothing. To summarise: a value of 0.5% in the above countries is the best we can hope for, as long as science cannot come up with any new tricks to boost the efficiency.

What does this value mean for the energy supply situation? Suppose we grow plants, harvest them and grow new plants, in a continuous cycle so that we create a sustainable energy source. Let us forget for a moment that we would have to use fertilizer to prevent soil depletion. How much land would we need? A back-of-the-envelope calculation is easily made. Let's take a (realistic) photosynthesis efficiency of 0.5%. Using the figures we saw earlier for the total energy use in the Netherlands (page 65) and assuming an average solar energy irradiance of 110 W/m^2, we come up with a surface area of almost 200,000 km^2. This is about five times the total area of the Netherlands.

Conclusion 1: *If we were to base present energy use in the Netherlands on the growing of plants in a sustainable way, we would need the entire Netherlands plus Belgium plus part of Northern France.*

We have not yet taken into account the energy that goes into plowing, sawing, harvesting, etc. and – in the case of wanting to convert our harvest into solar fuel – the energy used in converting it to alcohol or biodiesel. In unfavourable cases, the net yield may even be negative. In any case, we cannot expect the net yield to exceed the 0.5% mentioned above.

Let us turn to actual situations and consider the yield in three cases: sugar beets (one of the highest-yield crops in a moderate climate), potatoes, and rapeseed (which has become important in Europe). Also, let us not only use the alcohol (ethanol) that results

The efficiency of photosynthesis

(Source: H. de Groot, Leiden University)

In the photosynthetic process, water is used as a source of electrons,

$$2H_2O \xrightarrow{h\nu} O_2 + 4H^+ + 4e^-$$

and atmospheric CO_2 as a source of carbon,

$$CO_2 + 4H^+ + 4e^- \xrightarrow{h\nu} (H_2CO) + H_2O$$

where $h\nu$ is the energy of a photon taken from solar radiation (h is Planck's constant). Unfortunately, we lose part of the energy in the storage process. The light energy on the chlorophyll has a very short lifespan, about 10^{-9} s. When light energy is not immediately removed from the chlorophyll, it disappears again in the form of fluorescence emission: use it or lose it!
The two chemical reactions involved in photosynthesis given above require 8 photons and take about 1 ms, which is six orders of magnitude slower than the decay of the light energy on the chlorophyll. Since the organic material that stores the energy is in chemical equilibrium with the chlorophyll, this equilibrium needs to be shifted towards the fuel side to prevent fluorescence loss, while also sacrificing some of the energy. In practice, to drive a water-splitting reaction at 1.23 V, a minimum voltage of 1.8 V is needed, which can only be produced with chlorophylls absorbing light at wavelengths shorter than 700 nm. Because of the losses in the absorption of the light by the chlorophyll and the storage reactions, the maximum theoretical energy efficiency of photosynthesis is 12%, which is about half the theoretical limit for converting solar energy to electricity without storage. The storage losses depend on the desired storage time. Electricity is the most efficient way of using solar energy, since it can be transported at almost the speed of light, and there is no need to store it. The maximum time that solar energy can be stored in a chemical network is about 10^{22} s for thermodynamic reasons. However, this is more than long enough, considering that the remaining lifetime of the sun is on the order of 10^{17}–10^{18} s.
The conversion efficiency in the primary reactions of photosynthesis is very high at almost 100%. But there are some losses, for example, because the CO_2 concentration in the air is too small to be optimal (although the man-made increase reduces this problem, which is the one positive aspect of the global warming issue). This brings the energy efficiency down to about 6%. This is indeed achieved in healthy green leaves at low light intensities. But as soon as there is a lot of light, many plants exhibit saturation effects: the plant is unable to absorb all the light that it is offered. Moreover, the efficiency is reduced as a result of the respiration losses of the plant. These are, depending on the circumstances, circa 40%.
To summarise, the efficiency of photosynthesis under optimum conditions is about 3%. In practice, this is achieved in a favourable climate, by plants such as sugar cane, which do not suffer greatly from the effects of saturation. Most plants restrict the overproduction of biomass. The efficiency is then much lower, less than 0.5%, and obviously depends on the type of plant, climate, soil, use of fertilizer, etc. Finally, the purification and separation methods of solar fuels are far from optimal. Thus, there is still ample room for improvement.

from the fermentation of the sugar produced, but let us also include the fuel recovered from leaves and other organic 'waste'. We consider 'net' yield, by subtracting the energy used in the whole process. For sugar beets, for example, the yield per hectare (10^4 m^2) turns out to be 123 GJ of ethanol from the sugar and 91 GJ in the form of biogas from organic waste.
In total, this is 214 GJ. But since 57 GJ were used in the process, the net yield is 157 GJ/ha. The net yields for sugar beets, potatoes and rapeseed are:

Solar fuels derived from Dutch agriculture
(From: *Groenboek Energietransitie*, Senternovem, 2007)

	Net yield per year (including organic waste) in GJ/ha	in litres of oil per m^2
Sugar beets	157	0.42
Potatoes	73	0.20
Rapeseed	57	0.15

We can compare this to the amount of solar energy that is annually delivered to the soil in countries like the UK and the Netherlands. This is the equivalent of 90 litres of oil per square metre. This means that the net efficiency of sugar beets is, indeed, 0.5%, which matches our theoretical figures calculated above.

Conclusion 2: *If in countries like the the UK and the Netherlands solar energy is harvested and then converted into solar fuels, the of 90-mm layer of oil equivalent delivered by the sun annually is reduced to less than 0.5 mm.*

So, solar fuels are beautiful. They may very well contribute to the energy supply, especially as 'second generation' solar fuels (see page 99), but, from our figures, it is clear that expectations should not be set too high. We cannot expect to 'harvest' enough fuel in our small backyards to drive our car for a year, for which we need about 1,500 litres. The production of this amount of solar fuel requires approximately 5,000 m^2 of land.

Sunny countries

The prospects are more promising in countries with a more favourable climate and a higher solar energy irradiance. In these cases, we have the best of both worlds: we can harvest a higher percentage of a larger amount of energy. An example is sugar cane in Brazil. As we mentioned earlier, the efficiency of photosynthesis can reach about 3% in warmer countries. If we assume that the *net* yield is 2%, we end up with four times more energy produced than from sugar beets in a moderate climate. Since the solar irradiance is higher by another factor of 2, the net yield could be 8 times as high. This gives us about 3 litres of oil equivalent per square metre per year. Note that this is, however, restricted to a small climate zone.

Another example is palm oil. In well-suited areas (within 15 degrees latitude of the equator) the annual harvest is 0.5 litre per m^2. This is much higher than rapeseed in the Northern countries, but not much better than the solar fuel yield from Dutch sugar beets (see table p. 98), even though the solar irradiance is higher (see the figure on page 95). This type of yield would correspond to an average photosynthesis efficiency of 0.3%. This is not impressive. Moreover, these palm oil plantations should not be developed at the expense of rain forests, which is what is currently happening in places like in Southeast Asia, for example. If it does occur, it will further enhance global warming, since the original rain forest is a larger carbon reservoir than the plantations that replace it. We should keep in mind that it is the *reservoirs* that count, if we want to keep the extra CO_2 out of the atmosphere.

Solar fuel versus food

First-generation solar fuels, as they are called, are produced from numerous agricultural products that are also used for food. Biodiesel is produced from rapeseed, palm oil or soybean oil. Ethanol (alcohol) for mixing with petrol is produced from sugar, corn, and wheat through fermentation. The question is then whether this jeopardises food production. The steep rise in food prices after 2005 was partly as a result of the conversion of fertile agricultural fields for the production of solar fuels instead of food.

There are also chain-reaction effects. Here is an example from the US (M. Grunwald, *Time*, April 14, 2008). Around 2008, about 20% of corn production in the US went to the production of ethanol. This increased the price of corn. Eager to cash in, many US soybean growers switched from soybeans to corn, because it was more profitable. This reduced the soybean supplies, resulting in higher soybean prices on the global market. To meet global demand, Brazilian farmers then began producing more soybeans, on land that had previously been used for cattle pastures. In turn, cattle ranchers cleared more new pastures for grazing by cutting trees in the Amazon rain forest and the Cerrado savanna, ultimately releasing carbon into the atmosphere. So, in the end, Brazilian rain forest territory was destroyed by corn production for solar fuels in the US. This resulted in increased CO_2 emissions, as well as releasing relatively large quantities of laughing gas (N_2O) in the process of fertilizing the newly reclaimed pastures.

The good news is that the agricultural yield in many countries can be increased dramatically by improving crop selection and by using more efficient agricultural methods. This has been the case, for example, in many Eastern European countries. So, if efficiency improvements are larger than the growth in demand for food (which seems unlikely), there may be some future for growing such first generation solar fuels. But, the production of solar fuels still raises the ethical question of whether it is justified to 'put food for the poor into the cars of the rich'. Apart from possible food shortages, the increased harvesting of bio-ethanol crops and similar products to produce car fuel may create an interesting dilemma: shall we drink another glass or drive another 300 metres?

Second-generation solar fuels are less of a threat to food production. They are produced from straw, grass or other agricultural waste. These can be converted into fuel by the workings of yeasts and enzymes, but also by means of various thermo-chemical processes.

Research and development on this topic are in full swing. On an industrial scale, these processes may become available around 2020.

Obviously, one can also simply *burn* organic material, especially if it is reasonably dry (otherwise, a sizable percentage of the heat produced during combustion will be wasted on the evaporation of the water content). The burning of organic material is already being done on a large scale in some power plants.

But no major short-term contribution is expected from 'second-generation solar fuels'. The reason is, basically, that the current efficiency of photosynthesis is low, as outlined above. An illustration of the net agricultural yield in a moderate climate was presented in the table on page 98.

So shouldn't we use *infertile* soil, which is useless for food production anyway? Think of trees or grass, for example. This sounds like an elegant solution. However, we should realise that such natural systems are, in principle, closed cycles. All nutritious substances remain in the soil, so there is no need to add fertilizer. Once we start harvesting, the soil becomes depleted and the efficiency goes down. For example, experiments were done on prairie grass as a potential solar fuel on terrible soil in Minnesota, US, by David Tilman. Prairie grass does not compete with food production, and it requires no pesticides or fertilizer.

While this sounds good on paper, in practice it may be less of a success story. The harvest corresponded to the equivalent of 0.2 litres of oil per square metre, or an efficiency of 0.1 %. In addition, the harvest was not actually removed from the land, but burned on the spot. By doing so, the nutrients were recycled back into the land, so in a way fertilizer was used after all (see: David Tilman, *Science*, December 8, 2006, and Michael Russelle, *Science*, June 15, 2007). Consequently, this does not show how the soil will behave in the long term if the harvest is removed and used as raw material for solar fuels.

Meanwhile, the Jatropha plant is a major candidate for the production of solar fuels in tropical regions. It grows even on dry and infertile soil, so it does not have to compete with food production. It is poisonous and thus not suited for human consumption. Its seeds are 45% oil, which is well suited as solar fuel. The yield is 0.1 litre/m^2 per year and thus comparable to rapeseed in a moderate climate. Higher yields are possible, but only if extra water and fertilizer are used. So, the high yields that are sometimes claimed are not compatible with the use of dry and infertile soil. Even so, small-scale Jatropha production may be a solution for the energy demand of developing countries. It is not a solution for the enormous energy needs in richer countries, however.

New developments: algae and bacteria

As we have seen above, the problem with natural photosynthesis is its low efficiency levels: only a small part of the solar energy flux is actually being converted. Photovoltaic cells, which convert more than 10% for current commercial cells and possibly over 40% in the long term (see page 106) are much more efficient. Furthermore, they won't compete with food since they don't require fertile soil.

Can't we improve the efficiency of photosynthesis to the level of photovoltaic cells? To achieve this, we need to leave the trodden path of ordinary vegetation. Once we do that, scientific breakthroughs can open up new possibilities in this field. For example, experiments are being done with micro-organisms like micro-algae and cyanobacteria. There are at least two interesting developments in this field:
- *Converting CO_2 into carbohydrates*. We are familiar with algae that convert water and CO_2 into oxygen and glucose or other carbohydrates. This already yields a usable solar fuel. However, a much higher conversion efficiency can be reached for micro-algae. They achieve a photosynthesis efficiency of 5%, without being genetically optimised. They can also produce large quantities of triacylglycerol, the starting point for making biodiesel.
- *Converting water into hydrogen*. Some green algae and cyanobacteria can produce hydrogen from water directly, using sunlight, which is referred to as 'photobiological hydrogen'. This means a direct conversion of solar energy into fuel, whereas current solar fuel production uses solid organic material as an intermediate step. Efficiencies of about 2% have already been achieved. This provides a basis for improvements in efficiency through genetic engineering, as well as for suitable artificial systems. Following this road, it seems possible to achieve efficiencies of 7 to 10% in the foreseeable future, at solar irradiances of approximately 250 W/m^2.

This kind of 'artificial leaf' may be capable of providing a reasonable portion of our liquid fuels by the year 2050, if there is sufficient progress in science and technology. Ultimately, the efficiency of this *'third generation'* bio-energy may go as high as 42% (Source: H. de Groot, Leiden University and R. Rabbinge, Wageningen UR).

Such experiments are already being carried out on a small scale in the laboratory. To become meaningful for the energy supply, these experiments ultimately need to be performed on a larger scale without using fertile land, on, for example, open water. The 'production' of such systems could actually be done by nature itself. This may well be a fast process: just think of the speed at which algae grow during the summer. But here we should also not expect miracles. It will take a great deal of effort to realise these types of systems.

B2 | Solar Heat: Passive Use and Solar Collectors

The simplest way of utilising solar energy is certainly not the worst: use the solar heat *directly*: for home heating, or for heating an outdoor pool, for example. In these cases, which are often referred to as *passive use* of solar energy, the efficiency is quite high. Let's look at an example. If the sun is high and it is shining on a swimming pool, more than 90% of the solar energy ends up in the water: at normal incidence, the reflection is only 2%. If the pool is deep, or if the bottom is dark, almost no energy will be reflected back, so the efficiency will be over 90%.

There is a similar situation for home heating using windows. If the sun is low, a single-pane window will reflect only 8% of incoming radiation (4% in the front and 4% in the back). For a double-pane window this is about 15%. If we ignore the small part that is absorbed by the glass, the rest of the radiation ends up inside the home.

And almost all of this energy contributes to heating the home. Only the part that is reflected by light-coloured parts of the interior will escape back out through the window. This means that the 'efficiency' of this basic solar collector is circa 80%. This is the case even if the sun is higher, up to approximately 50°. Beyond that angle reflection becomes significant. But for moderate latitudes this happens only during the summer, when we don't need to heat our homes anyway.

Exactly how much energy windows and greenhouses collect throughout the year can be seen in the figure presented earlier on page 94, where the solar irradiance is given for various angles.

Solar collectors are simple in principle. A black plate at ambient temperature will absorb almost all incident solar radiation, so its efficiency is nearly 100%. However, when it starts heating up, losses begin to play a role. Part of the absorbed heat is re-radiated, or lost to the ambient air by conduction. So the efficiency goes down as the temperature increases. A solar collector that has to produce very hot water will have a lower efficiency than one which produces lukewarm water. Therefore, solar collectors are the most profitable in situations that require relatively low temperatures, like swimming pools or a heat reservoir for heat pumps. The efficiency here can easily be as high as 70% in these cases.

- When applied to hot-water taps, the efficiency is lower: around 40%. If we want a year-round, solar hot-water supply, the collector would have to be quite large to adequately satisfy needs even in the winter, but this would be much larger than necessary during the summer months. In a moderate climate, this type of 'solar boiler' with a collector surface of 4 m^2 and a buffer volume of about 150 litres will supply two-thirds of the hot water demand over the course of a year. The rest must then be generated by a back-up heating system that is powered by electricity, oil or natural gas. Obviously, the precise choice of these numbers is always an economic trade-off.

- When applied to home heating, the efficiency may be comparable. However, we now face the problem that the collector's yield is the smallest when we need it the most – during the wintertime. This, of course, means having a stand-by unit, as noted above, powered by natural gas, for example. How this actually works out in practice is illustrated in the figure on p. 103.

If we wish to take full advantage of the summer sun using a solar collector for our home heating, we have to figure a way to store the heat for a period of several months. This type of seasonal heat storage system requires many cubic metres of water or soil, which is heated up during the summer and acts as a heat reservoir during the winter. In such cases, large units are more attractive than smaller ones. The reason is that the stored energy is proportional to the *volume* (proportional to the dimension cubed) while the heat losses are proportional to the outer surface *area* (proportional to the dimension squared). In other words: if we choose a certain geometry (a cube, for example) twice as large a unit will store eight times as much heat and will have only four times the losses. This makes this type of storage attractive for houses that are lumped together such as housing blocks.

Developments are under way for chemical storage systems, which do not have the thermal loss problems when storing heat from one season to another. These 'thermo-chemical materials' are based on chemical reactions between two components, which are kept apart during storage time, but which produce heat when brought together. In the reverse process, the two components can be separated by heat, thus closing the cycle. Schematically:

$$AB + \text{heat} \leftrightarrow A + B$$

In view of the enormous quantities of heat that need to be stored for seasonal heat storage, it is not obvious that these systems will indeed be used on a large scale in the future.

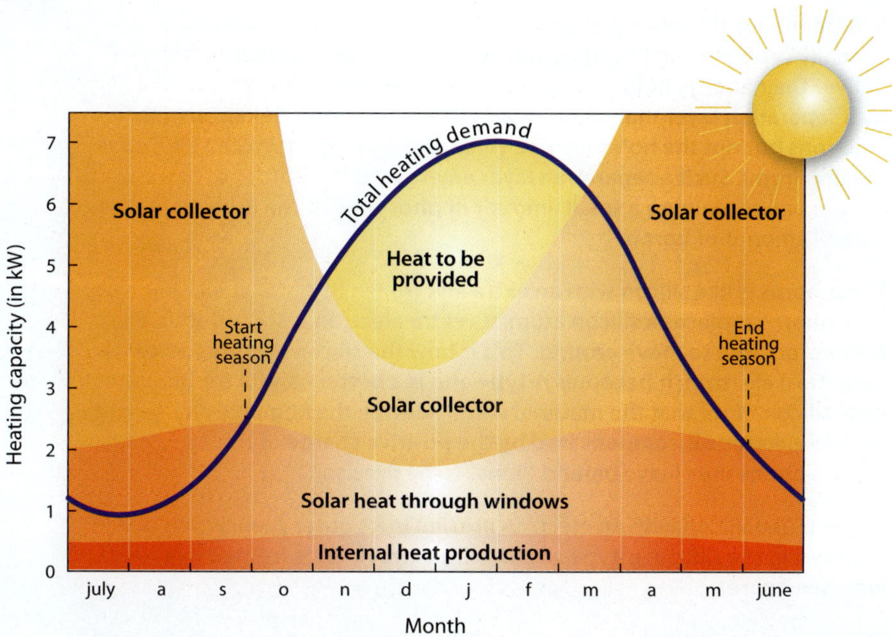

Schematic example of the annual heating balance of a home equipped with solar panels for heating (orange area). Red areas represent internal heat production and the contribution of windows, respectively. Since the house does not have long-term storage the total heat production during the winter is not enough to heat the house. As a result, the yellow area has to be supplied by a stand-by unit.

B3 | Electricity by Photovoltaic Cells (PV)

Silicon-based cells

Photovoltaic (PV) cells convert incoming solar radiation directly into electricity. The current cells (based on silicon) have efficiencies up to 25% in the laboratory and 14–20% in commercial production. The operating principles are outlined in the box 'How does a solar cell work?'.

How does a solar cell work?

The most commonly used solar cell is based on crystalline silicon. In pure form this is a semiconductor which has very few free electrons in the absence of chemical additions or light. In that case, the electric resistance is high. However, by absorbing sunlight (in the case of silicon: photons having an energy of more than 1.1 eV) free electrons can be produced. This yields mobile negative electrons and mobile positive 'holes'. Each electron-hole pair represents a potential energy of 1.1 eV (if the photon has more energy, the rest is lost as heat).

In order to use this potential energy, the electron and the hole must be spatially separated so that they can flow through an external circuit and deliver their energy before they are allowed to recombine. This is achieved by a separation layer that contains an electric field. This field transmits the electrons but not the holes (or vice versa, depending on which side you're looking from). Such a separation layer emerges at the border of two types of silicon: one containing a small amount of phosphorus, the other containing a small amount of boron.

Phosphorus is like silicon with an extra electron in its outer shell. When phosphorus replaces a silicon atom, its extra electron is superfluous and it is therefore free to move around. This is how the silicon acquires an extra (negative) electron: it becomes n-type and is a better conductor than the pure silicon. Note that the material is still neutral: the negative charge of the mobile electrons is compensated by the positive charge of the fixed phosphorus atoms they leave behind ('ions', to be precise).

Boron is like silicon with an electron *missing* in its outer shell. When boron is added to the silicon, p-type silicon with extra holes is formed in a similar way (see figure).

Where n- and p-type silicon meet, mobile electrons and holes permanently cross the border, driven by the large differences in concentrations between both sides. While doing so, they leave behind the fixed charge at the phosphorus and boron ions. Therefore an electric field builds up which becomes just large enough to counteract the tendency of electrons and holes to cross the border as a result of the concentration differences. Note that the whole structure is still neutral because the amount of positive and negative charge on both sides of the border is equal.

If *extra* electrons and holes are now produced by absorption of incoming sunlight (which can happen on either side of the border; both in the neutral n- and p-type materials *and* in the interface layer) they will be separated by the electric field. This produces a voltage difference between the two sides of the cell. If current is allowed to flow, the *combination of voltage and current* produces an *electric power*; the aim of solar cell use.

How does a solar cell work?

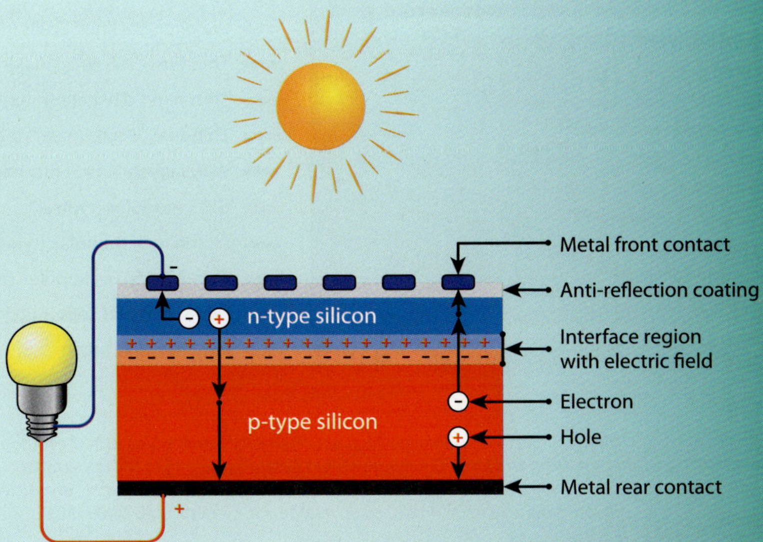

The schematic of a silicon solar cell. The blue and red regions are n-type and p-type silicon, respectively. The interface layer contains an electric field (see text) and separates light-generated electrons and holes, which are subsequently transported to the external circuit through electrodes attached to the top and the rear of the cell. An anti-reflection coating on top reduces reflection.

Why isn't the efficiency higher? Well, first of all, the efficiency (the fraction of the solar radiation that is converted) is limited because not all of the colours are equally effective in producing charged particles from neutral atoms.

In the silicon case, maximum efficiency is achieved only with radiation in the near-infrared (a wavelength of 1.1 micrometres). Light which has a longer wavelength does not contribute at all, since the corresponding photon energy is insufficient. Light which has a shorter wavelength, has an excess of energy. This excess is released as heat.

Moreover, there are also losses that occur in the process of recombination of charges (see box 'How does a solar cell work?'). We can decrease these losses by concentrating the sunlight, which works well in regions that enjoy much *direct* sunlight. The bottom line is that the maximum efficiency that can be achieved by a single-junction cell at normal solar irradiance is slightly under 30%. For small surfaces in laboratory conditions, 25 to 26% can be achieved, which means that these cells are almost perfect. Commercial solar panels with large surfaces have a maximum efficiency of 20%. In practice, figures range from 8 to 15%.

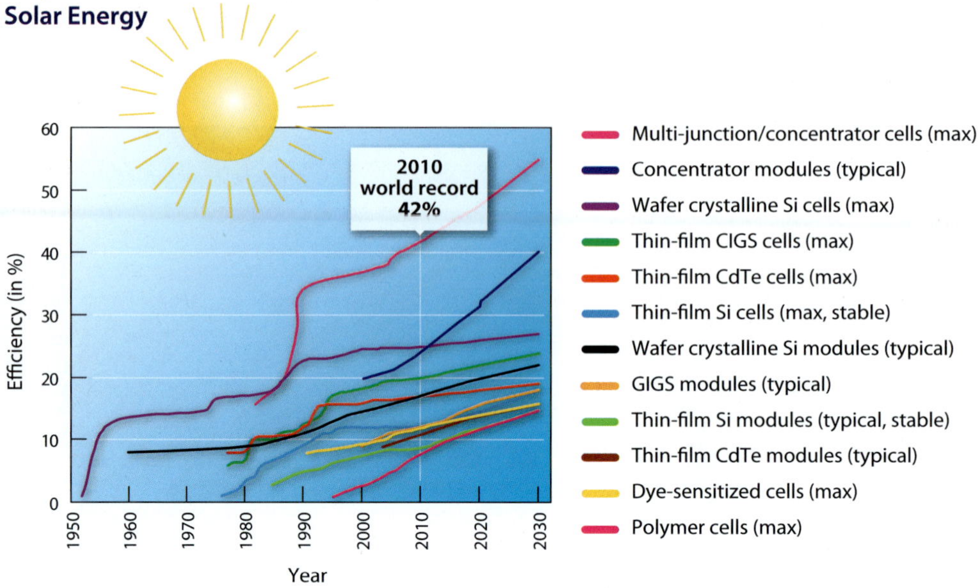

Efficiencies of various photovoltaic ('PV') cells over the years. High-efficiency cells used in the famous Nuna solar cars, winners of the World Solar Challenge race in Australia, belong to the uppermost category of multijunction/concentrator cells, which are also used in spacecraft. They have an efficiency of over 26%, but are still too expensive to be produced on a large scale (courtesy W. Sinke, ECN).

Other types of solar cells

If we want to achieve increased efficiency we can combine different materials that have different colour sensitivities: the light transmitted by one layer can be absorbed by the next one. They are called multi-junction cells. But there are more types of PV cells. Roughly speaking, we can distinguish four categories (Courtesy: W. Sinke, ECN):

- Crystalline silicon cells ('wafers'). These are by far the most commonly used PV cells.
- Thin-film cells based on silicon (amorphous or microcrystalline), copper-indium-diselenide or copper-indium-sulfide (CIS) and variations on this theme. Cells using indium will probably not be used on a large scale since this metal may become scarce fairly soon, in contrast to silicon.
- Multi-junction cells based on 'III-V'-semiconductors (the family around gallium arsenide).
- New, as yet not commercially available types. Examples include 'organic' solar cells based on polymers or a dye in combination with a transparent semiconductor like titanium oxide: *'dye-sensitized cells'* (see figure above).

Meanwhile, so-called PhotoVoltaicThermal (PVT) systems have also been developed. They combine electricity generation with heat production, by converting most of the energy not used for electricity into heat. The total efficiency level is therefore high.

Efficiencies

The above figure displays the efficiency of various types of solar cells over the years, including an extrapolation of the expected yield in the coming years. The efficiency record so far (as of 2010) is 42%, achieved using multi-junction/concentrator cells.

Commercial feasibility

The 'energy-payback-time' of a complete solar energy system is 3 years (2008, the Netherlands' climate). This means that the amount of *energy spent to make* the system is produced *by* the system within three years. This period corresponds to one-tenth of the technical life expectancy of the system (the warranty period of solar panels is sometimes as long as 25 years).

The payback-time for the *purchase price* of the system is much longer. This is as yet the most important barrier for large-scale use. This is evident from the price of solar electricity, which is 0.25 to 0.30 Euro/kWh in sunny countries like Spain, and 0.50 Euro/kWh in Northern countries like the Netherlands. However, expectations are that this will go down, possibly reaching one-tenth of these figures in the long run.

Some characteristic parameters for solar PV cells, and the expected future development (Data from W. Sinke, ECN)

	1980	2007	2015	2030	long term
Price complete system in €/W_p (excl. VAT)	> 30	5	2.5	1	0.5
Production costs of electricity, in 2006 €/kWh	> 3	0.50	0.25	0.10	0.05
Typical efficiency of commercial modules, in %	< 8	≤ 15	≤ 20	≤ 25	≤ 40
Energy-payback time of complete system, in years	> 10	3	1.5	< 1	< 0.5

The table above lists a few characteristic figures for current solar-cell technology in the Dutch climate. Here the number of 'peakwatts' (W_p) is the electric power delivered by a solar panel in full sunshine, which amounts to 1,000 W/m^2. In terms of an average yield in a moderate climate like in the Netherlands, this means that one peakwatt corresponds to approximately 0.8 kWh per year (or, on a continuous basis, to 0.09 watt as a day-and-night average over the year).

Surface area needed
In addition to producing a high-quality form of energy (*i.e.*, electricity) solar cells have much higher efficiencies than photosynthesis. Even so, large surfaces are still needed to produce substantial amounts of electricity.

Let's take a power plant that produces 1,000 MW of electricity as an example. Suppose we want to generate this amount of electricity using solar cells, assuming a somewhat optimistic efficiency of 15%. Given the night-and-day, year-round average solar irradiance of 110 W/m^2, this would require a net solar-cell surface of 60 km^2.

This is quite a lot. On the other hand, a multiple of this area is available in the form of roofs and façades that are unutilised, comprising a few hundred km^2. For densely populated countries, therefore, small-scale decentralised use of solar cells seems more attractive.

Now, suppose we wanted to generate our own domestic electricity using PV panels. How large a surface would we need?

Let us assume that the average electricity use of a household is 400 W on a continuous basis (which is the case for countries like the Netherlands). The sun delivers, on average, about 110 W/m^2 on a horizontal surface, and slightly more if we place the panels at an angle (see the figure on page 94). Again, let us take a slightly optimistic efficiency level of 15%. This means that we would 'harvest' about 18 W/m^2 on a continuous basis. So, to generate 400 watts, we would need a cell surface of slightly more than 20 m^2. This is the surface area of a large garage roof.

The price follows from the table on p. 107 if we know the required number of 'peakwatts'. Since we need an average of 400 watts and 1 peakwatt corresponds to 0.09 watt average, we need 4,400 peakwatts. If we assume € 5 per peakwatt, we arrive at € 22,000. This is approximately the price of the car parked *under* the garage roof. But the price is falling rapidly (while costs of electricity from the utility company will probably go up).

We have neglected one important aspect. The sun does not always shine when we need the electricity. So, we have a storage problem. This will be addressed extensively in chapter 12.

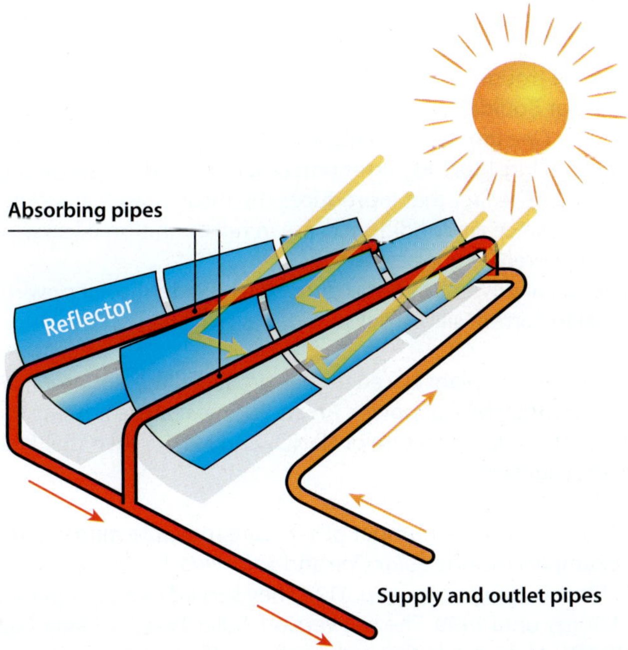

Schematic of a reflecting parabolic trough for concentrating the sun's rays

B4 | Electricity by Concentrated Solar Power (CSP)

In regions where much of the solar energy comes from cloudless skies, we can concentrate the sun's rays using mirrors; this is called Concentrated Solar Power. For example, we can heat a steam boiler that, in turn, drives a turbine. The sun in this case replaces the traditional fire. Or we can use the focused rays of the sun to drive a 'Stirling engine' (or hot air motor, see p. 110). In either case, their efficiency may surpass that of the solar PV cells.

Although the CSP approach and PV cells share the problem of a sudden cloud obscuring the sun, the CSP approach can more easily cope with this problem by temporarily storing the heat. See, for example, the Andasol 1, page 110.

In CSP systems, we obviously need some mechanism to follow the sun during the day. Exactly how complicated that is depends on the type.

When using a ***parabolic trough*** the focal point becomes a focal line. In that 'focal line' one puts tubes filled with oil that is heated to high temperature by the sun. In turn, the oil heats water driving a steam turbine. The nice thing is that the construction is relatively straightforward. In addition, the 'two-dimensional tracking' needed for parabolic mirrors to follow the sun is now replaced by 'one-dimensional tracking', which simplifies construction considerably.

A whole field of these troughs can be combined into one large facility. A prototype was built in 2007, near Boulder City, Nevada, US, about 30 km outside of Las Vegas: ACCIONA's *Nevada Solar One* (not to be confused with Solar One in California, see below). It has 760 troughs totaling 180,000 mirror reflectors (see aerial view on page 111) and has a power of 64 MW. The fact that it produces the most energy at around midday is handy in that climate, since air-conditioners are in full operation everywhere during this time. Storage is therefore unnecessary.
The efficiency (electrical power delivered divided by incident solar power) is around 15%, roughly equal to current silicon cells.

In Andalusia, Spain, another plant based on this principle, the *Andasol 1*, was built in 2008. It produces about 50 MW of power. An additional feature is heat storage for about 7 hours in molten salt (a mixture of $NaNO_3$ and KNO_3), such that it can deliver electricity even after sunset.

A Solar Power Tower has a large number of separate movable mirrors that are focused on one 'boiler'. Examples include *Solar One* and *Solar Two*, located in the Mojave desert near Barstow, California (see photo on p. 112). They served as a pilot plant for the US Department of Energy until 1999. The last version, Solar Two, contained 82,750 mirrors and produced 10 MW of power. In this case molten salt was also used as a storage medium.
Solar Tres, which is similar to Solar Two was built In Andalusia, Spain and produces 15 MW of power with a total mirror surface area of 240,000 m^2. Molten salt is also employed here as a high-temperature storage medium. It stores 600 MWh of energy, so that the plant should be able to produce electricity continuously during the summer months.

Stirling engines ('hot-air motors') can be used to generate solar energy. A Stirling engine operates on a principle quite different from the ordinary internal combustion engine. In an ordinary engine, the pistons are pushed away by an igniting air-fuel mixture. In a Stirling engine, a closed volume is heated *externally*, which can be done by any heat source including the sun. The principle is that some gas (for instance, ordinary air, or helium) is compressed while it is cold, then – while it is still compressed – moved to a hot region where it is allowed to expand. Finally the gas is moved to the cold region again and the cycle is repeated. The key is that the expansion of the hot gas produces more mechanical energy than is needed to compress the cold gas. So, in such a compression and expansion cycle, net mechanical energy is produced from heat with a reasonably high level of efficiency.

In the United States, Sandia National Laboratories has built a pilot plant consisting of six small solar-energy units equipped with Stirling engines (see picture on page 112). Each Stirling engine is placed in the focal point of a disc measuring 11 metres in diameter, and produces 25 kW. This is equivalent to the power of a small car. If we do a back-of-an-envelope calculation, using a solar irradiance of 1,000 W/m^2, we come up with an efficiency of 26%. A record efficiency of 31.25% was recorded on January 31, 2008.

So, Stirling engines seem perfectly suited for these moderate energy levels.

The parabolic-trough solar power plant ACCIONA's Nevada Solar One, near Boulder City, Nevada, US

The first European commercial solar tower (2007): the PS10 near Sevilla, which generates 11 MW of power

The Solar One in the Mojave Desert in California, with a peak power of 10 MW

Six small solar power plants using Stirling engines at Sandia National Laboratories in Albuquerque, New Mexico, US
They set a world record in 2008 by achieving a net efficiency of 31%

Solar car

How realistic is it to have cars powered by solar energy?
Let's do a back-of-the-envelope calculation here. As we have seen above, the maximum solar irradiance anywhere on earth is 1,000 W/m^2. If we use current silicon solar cells with an efficiency of 15%, and if we assume a cell surface of 5 m^2, we end up with electric power totaling 750 watts or 0.75 kW. That is one 'horse power' or the power of, say, a vacuum cleaner. This is not much compared to present-day car engines, which generate about 50 kW or more. Driving a car purely on solar power thus seems impossible.

But how did those people in the World Solar Challenge in Australia manage to drive over 100 km/h on solar energy? First, let's point out that this vehicle wasn't a normal car at all. It is an ultra-light and ultra-streamlined one-person vehicle. In addition, the solar cells are space-technology cells with twice the efficiency of current commercial silicon cells. Take the winning Nuna 4, a 2007 vehicle designed by Delft University of Technology. Using those super solar cells and an area of 6 m^2 the car was powered by about 1.5 kW (or 2 hp). Its mass was only 190 kg, its rolling resistance was only 1/10 and its drag only 1/6 of that of a normal car.

But what about technological breakthroughs? Can we expect to be driving a car much like our present car on solar energy in some distant future? Unfortunately, the answer is no, even if we develop solar cells that achieve 100% efficiency. That is, if we assume that we do not have some hidden energy stored somewhere, and use only the solar power radiated on the spot. Remember the 1 kW per m^2 that the sun delivers? Assuming an optimistic 5 m^2 of solar cells, we end up at 5 kW at best. If we were to drive the model car featured on page 52, we would not be able to exceed a speed of 60 km/h. And if we consider the possibility of a cloud appearing at any moment, we have no choice but to use some kind of storage system.
For more on storage systems, see chapter 12.

The whole world on solar electricity?

Wouldn't that be a great idea, to have the whole world using electricity generated by solar energy? Let us just estimate how much it would take. Total annual, global electricity consumption was already provided in chapter 6. Converted into an average on a continuous basis this translates into 2,000 GW, or 2,000 large power plants in continuous operation. Let us look ahead, and assume that this will double to 4,000 GW by the time solar energy technology begins to mature.

If we want generate solar electricity, it would be wise to do it in favourable locations. As we can see from the 'globe' on page 95 we can easily locate spots where the solar irradiance is relatively high: around 250 W/m^2 on average.
In this sense we're lucky because most of these locations happen to be in deserts, which are not of much use anyway in terms of agriculture, etc. Take the Sahara, the Gobi desert, the Mojave desert; there are plenty to choose from.

Now, we have two options: we can take the PV-cell route or the CSP route. Either way, the efficiency level will be around 20% (which is also a reasonable estimate for PV cells, given the progress being made). This means that we can expect an electricity generation of 50 W/m^2 on average, or 50 MW/km^2. For the 4,000 GW quoted above we would need 80,000 km^2. This is twice the area of the Netherlands located in the Sahara desert.
An area of this size is certainly available in the desert, but covering it completely with solar cells or mirrors may not be so simple. Remember, again, that this is just *electricity* production, which represents roughly only one-third of *total* energy used.

It sounds like a nice plan, a huge power plant in the Sahara desert. What are the drawbacks? For one thing, we should not forget the day-and-night cycles. This requires some sort of storage. In addition, there is the need to transport this energy over large distances. Both of these issues point in favour of decentralised power plants, *i.e.*, generating electricity at the location where it will be consumed. Granted, this may bring us to locations with less favourable conditions as far as solar influx is concerned, but if the efficiency of PV cells continues to improve and the cost keeps going down, this may end up still being feasible.

In summary: if sufficient progress in science and engineering continues to be made, a combination of centralised and decentralised electricity generation using solar power should eventually become feasible. We should keep in mind, however, that the most likely long-term solution will be a hybrid of various technologies of which solar electricity will be just one.

The whole world on solar electricity?

Conclusions regarding Solar Energy

- Whenever the sun is high, its irradiance is almost exactly 1,000 watts per square metre taken at a perpendicular to the sun's rays.
- The day-and-night average over a year of solar irradiance hitting a horizontal surface is globally on average about 160 W/m^2. Expressed in terms of oil, this means an annual harvest of about 135 litres per square metre. In a climate like that of the Netherlands these figures are 110 W/m^2 and 90 litres per square metre, respectively.
- Plants and trees do not grow very fast. The annual average net efficiency of photosynthesis in countries with a moderate climate is only about 0.5%. This corresponds to a yearly harvest of 0.5 litre of oil per m^2. In a more favourable climate, this may be larger by up to a factor of 5. The advantage of using photosynthesis as a conversion mechanism is that the process provides free storage .
- Solar fuel manufacturing from crops can jeopardise food production. It may even be counterproductive from a global-warming perspective. 'Second generation' solar fuels that employ non-edible organic material have better prospects. 'Third generation' solar fuels that rely on scientifically engineered bio-solar cells or artificial leaves may achieve 40% efficiency.
- In a moderate climate like that of the UK or the Netherlands a south-facing window yields its highest heating contributions during spring and autumn, which are circa 110 W/m^2 on a daily average.
- Solar collectors that collect heat have 40 to 70% efficiency, depending on the type of application. If they are used for hot water they can take advantage of the high solar influx during the summer months. If they are used for heating homes, long-term storage ('seasonal storage') is required. This is feasible primarily for large units like a housing block or office buildings.
- Solar cells (Photovoltaic or PV cells) for large-scale use have efficiency levels up to 15%. The electricity they produce is still expensive: € 0.25/kWh in sunny countries like Spain and up to €0.50/kWh in countries with a moderate climate.
- In sunnier areas of the world, there is a good alternative to PV cells, called Concentrated Solar Power (CSP). Several pilot plants have already been built, in the US and Spain, for instance. Efficiency levels are between 15 and 30%.
- The total world electricity consumption at present could be supplied by a 40,000-km^2 area in the desert, roughly the area of the Netherlands. However, in view of storage and transport logistics, it seems more viable to also use more decentralised solar units.

9 | Wind Power

Wind power is basically solar power in disguise. After all, wind is caused by the unequal heating of the world's surface by the sun. The immediate conclusion is that the global potential of wind energy is only a fraction of the potential of solar energy. The same is true, by the way, for hydropower, and for wave energy, both of which are driven by the weather, which, in turn, is dependent on solar energy. The exception is tidal energy (which is more like moon-related energy).

Nevertheless, the potential for wind energy is appreciable, and its advantage is that it delivers direct mechanical energy, which can be converted into electricity almost without any loss.

A | The Potential of Wind Power

Just how much energy does the wind represent? This very much depends on the location, as we will see further on. The crucial factor here is the wind speed, since wind power is proportional to wind speed cubed. Twice as high a wind speed means eight times more power. This may be understood as follows. The *force* exerted by wind on a surface is proportional to the speed squared, as we saw earlier when we were looking at the effect of drag on a car (page 51). But *power* is force times distance per second, or force times speed. And since force itself is proportional to speed squared, power is proportional to speed cubed.
The exact derivation of wind power is provided in the box 'The power of wind'.

So, if the wind speed is v, the wind power is proportional to v^3. This is a vital aspect of wind energy. It means that wind turbines should be placed where the wind speed is reasonably high, for example, along the coast or at sea. Furthermore, we can use the fact that wind speed increases with height. Wind turbines are, therefore, best placed at a higher elevation. Moreover, large (and thus tall) wind turbines will be more profitable than smaller, shorter ones.

To find the best spots for the placement of wind turbines we should know average wind speeds. The figure on page 118 shows lines with equal average wind speeds across the Netherlands, measured at two different heights. We can see that, indeed, wind speed increases with height. Furthermore, we notice that a small windmill in the eastern part of the country may end up producing only one-eighth the power of a similar windmill placed along the north-western coast.

Something unexpected comes up at this point. It is not the average wind speed cubed that matters, but the average *of* the wind speed cubed. This may seem like semantics, but it can make a huge difference. This is because higher speeds have an unusually large impact in v^3. Let us look at an example.

The power of wind

The energy of a mass *m* of air moving at a speed *v* is its kinetic energy $\frac{1}{2} mv^2$. How large is the mass *m* that is passing through the rotor of a wind turbine per second? Let's call the area swept by the rotor *A* ($= \pi R^2$ with *R* its radius), and consider a space with a length of *v* behind the turbine, where *v* is the wind speed, or the distance covered per second by the moving air mass.

The volume *Av* thus contains the mass which passes through the turbine per second. That mass equals $Av \times \rho$, where ρ is the density of air, in terms of kg/m^3.

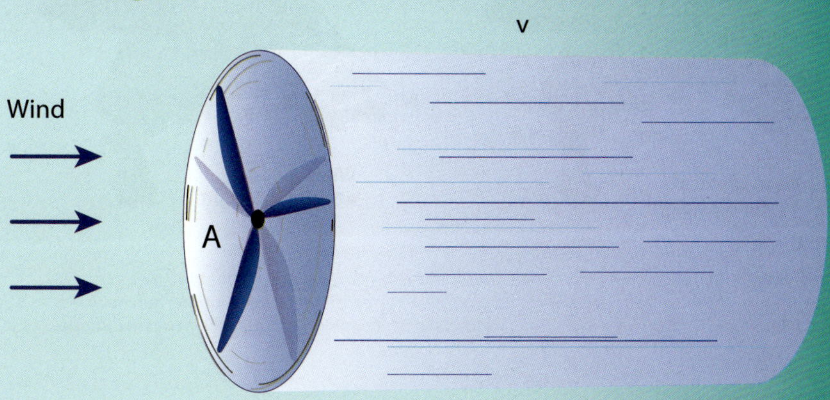

So, the energy that passes through the rotor per second is $P = \frac{1}{2} (Av \times \rho)v^2$, or

$$P = \frac{1}{2} A \rho v^3.$$

This is the power that the wind turbine would directly produce if it were extracting all of the available kinetic energy from the wind. However, this is impossible, since it would mean that the air speed would be reduced to zero, and all motion would stop. On the other hand, if we want to extract energy from the wind, we have to extract some of its speed. It turns out that the optimum is achieved when the wind speed is reduced to 1/3 of its original value. The corresponding power extracted from the wind turns out to be 16/27 of the power that was present in the moving air mass. This is known as Betz's theorem.

We can conclude that a wind turbine with a rotor with a radius *R* can produce a maximum power of

$$P_{max} = (16/27) \times \frac{1}{2} \rho v^3 \times \pi R^2 .$$

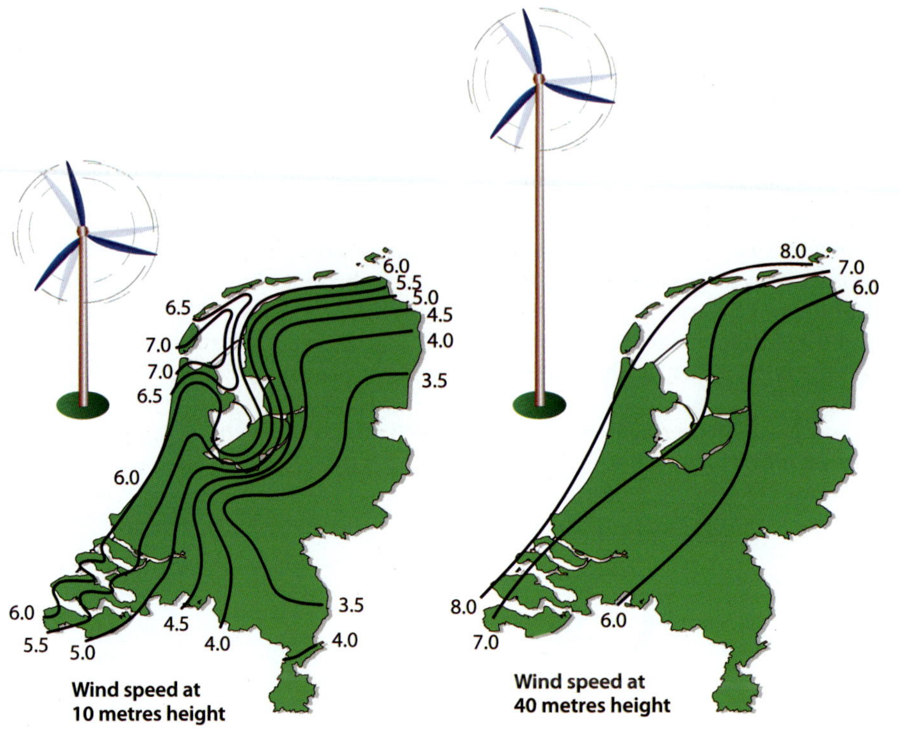

Annual average wind speeds in metres per second: on the left measured at a height of 10 metres and on the right at 40 m. The wind pattern is seen to closely follow the coast-line at low altitude, but much less so at higher altitude. Note also the increase in wind speed when going from 10 to 40 metres in altitude.

Suppose that one day we have a wind speed v of 10 m/s (which contributes 1,000 in terms of v^3), while the next day we measure 0 m/s (which contributes nothing), the average contribution is 500. In contrast, if, for two days, we have a constant speed of 5 m/s, which contribute 125 each in terms of v^3, the average of 125 is lower than the earlier two days by a factor of four.

If we take the situation in the Netherlands as an example, the annual average amount of wind power generated along the west coast by small wind turbines is about 100 W/m^2 while for large (and thus also taller) turbines, it can reach up to 200 W/m^2. The relevant number of square metres is the area that is being swept by the rotor blades. For example, a wind turbine with a rotor diameter of 50 m (sweeping an area of almost 2,000 m^2) will produce an annual average power of between 200 to 400 kW. This is mechanical energy, that can be converted to (almost) the same number of kilowatts in terms of electricity.

Note that this is not the *nominal* power, which is what is usually quoted for wind turbines. This nominal (or installed) power refers to the maximum power that the turbine's generator can produce in favourable wind conditions. The *average* power delivered is only 25 to 30% of the nominal power. This so-called capacity factor of 25 to 30% is often not taken into account, which causes a lot of confusion.

Why is a wind turbine's average production so much lower than its nominal value? Obviously, the basic reason is the large variability in wind speeds. If wind speeds were constant, then nominal and average output would be the same. The large discrepancy between the two is enhanced by the fact that the power is proportional to v^3, as seen above. A few peaks in wind speed have an enormous impact. But, since these occur so infrequently, it is economically not attractive to design a turbine that also uses the highest wind speeds. This is similar to a hotel, which does not have a sufficient number of rooms to accommodate every client on the busiest day of the year. So, the maximum power of the generator will not match the maximum wind power of those rare peak days. This places an upper limit on a wind turbine's capacity to generate power.

Not only is there a limit to the power, but there are also upper and lower limits to wind speeds used. In stormy weather, many wind turbines have to be taken out of opera-

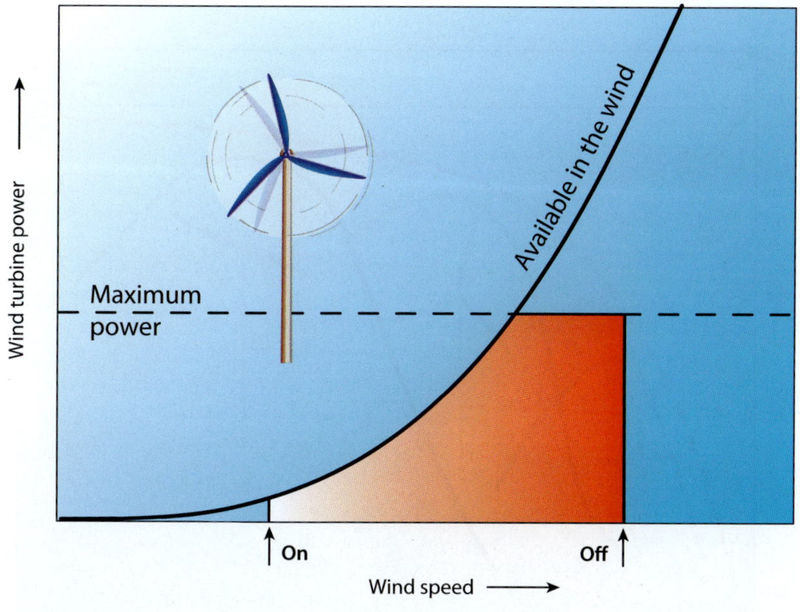

The available wind power (solid curve, proportional to v^3) and the power generated by a wind turbine as a function of wind speed. The turbine starts working at a minimum wind speed, the 'cut-in speed' (in the graph marked 'on'), while, at very high speeds, it reaches a 'cut-off speed', at which point many turbines will be taken out of operation to prevent damage to the turbines ('off'). Along the vertical axis, the output power has an upper limit determined by the generator ('maximum'). The red area between these limits represents effective production.

tion to prevent damage (although modern turbines can continue operating at a power level that is lower than the wind power). This puts an upper limit on the wind speeds employed for the generation of power. At very *low* wind speed the turbine won't even turn. Taking these restrictions into account, the effective operational power of a wind turbine is schematically represented by the red area in the graph above.

B | Types of Wind Turbines and Their Efficiencies

Is there much room for improvement in the wind turbine industry? The graph below shows performance figures for various types of wind turbines. The efficiency is shown as a function of rotor speed in terms of the ratio between the speed of the tip and the wind speed, the so-called 'tip-speed ratio'. The efficiency is defined here as the fraction of the available wind power that is actually extracted by the turbine, with 16/27 as the maximum based on Betz's theorem (see box 'The power of wind'). For each type of turbine the corresponding curve shows that the optimum is reached at a certain value of the tip-speed ratio. The red dashed line illustrates the trend toward the theoretical limit.

The efficiency decreases on either side of the optimum, where the rotor turns either faster or slower.

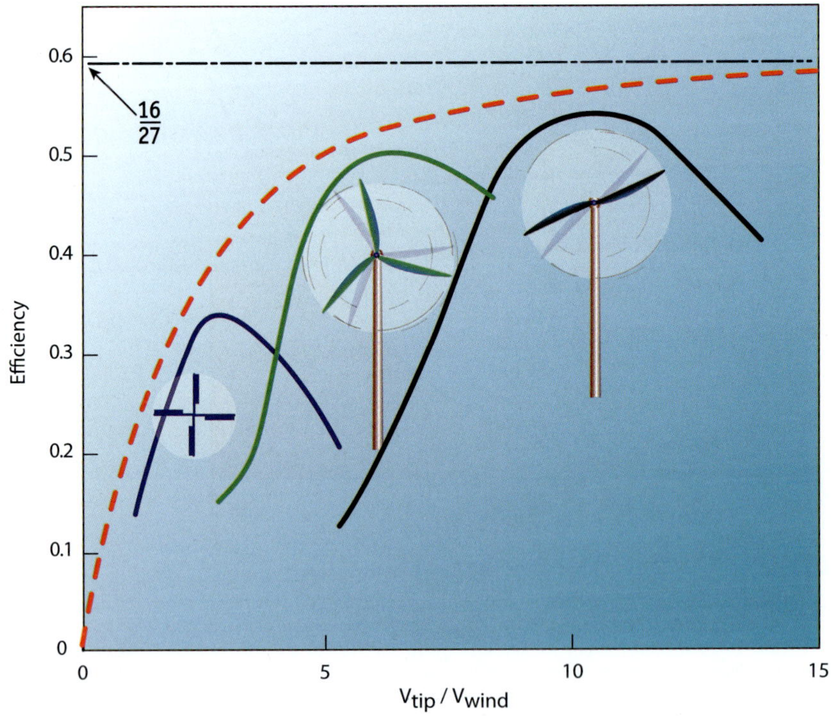

The efficiency of various types of wind turbines at increasing rotation speeds in terms of the ratio between tip speed and wind speed. Each turbine type has its own optimum, which is dependent on the number of rotor blades. The efficiency achieved by modern wind turbines almost reaches the theoretical limit of 16/27 (see box p. 117).

We notice four things here.

- First, traditional four-sail windmills perform pretty well. They reach about half of the theoretical limit.
- The more blades a turbine has, the faster the rotor will turn to reach maximum efficiency. This is logical, since the oncoming wind should be intercepted at certain regular intervals. So, two-blade turbines must turn faster than three-blade models.
- The maximum efficiency for each type of turbine is reached when the tip speed is a certain ratio of the wind speed (for example, the tip speed is 10 times the wind speed for two-blade rotors). For larger rotors, this tip speed is reached at fewer revolutions per minute. This means that larger rotors turn at a slower rate.
- Modern wind turbines perform close to the theoretical limit. This means that there is not much room for improvement.

It is remarkable that two-blade turbines actually perform slightly better than three-blade turbines. Why, then, do we see so many three-blade turbines (which are more expensive to build)? It is because of two societal arguments. First, wind turbines make noise, and the slower they turn, the less noise they make. This argument favours the three-blade models over two-blade models. Secondly, most people hate fast-moving rotors. The slower they turn, the less annoying they are. This argument also favours more blades, which reach their optimum at a lower speed as shown in the figure on p. 120.

These arguments play less of a role for wind farms located in the sea. So, we should not be surprised if we begin seeing two-blade turbines in future wind parks in the sea.

C | Space Demand and Wind Farms

The space demand of wind turbines

If we put a lot of wind turbines together at a windy spot they could replace a traditional power plant. However, they should not be placed to close to each other, since they would fall into one another's lee, as every sailor knows. This is true for *all* directions because they also influence each other sideways. The underlying reason is the behaviour of the wind profile. Remember that for the turbine to extract power from the wind, it has to slow down the moving air. So the wind speed around a turbine is slowed down. For optimum operation, the wind speed should be reduced to 1/3 of its original (see box 'The power of wind'). But this means that the cross-section of the wind profile must become three times as large to keep the air flowing. This may be compared to the behaviour of highway traffic approaching a toll booth: lower speeds mean larger flow areas. The behaviour of the wind profile is schematically shown in the figure on p. 122.

So, when wind turbines are placed too close to one another the efficiency declines. Just how close together can they be without jeopardising production? This depends on the situation. If wind turbines are placed in a row perpendicular to the prevailing wind direction (along the coast, for example) it is sufficient to position them 2 to 3 rotor diameters apart. However, turbines in a wind farm with neighbours all around require

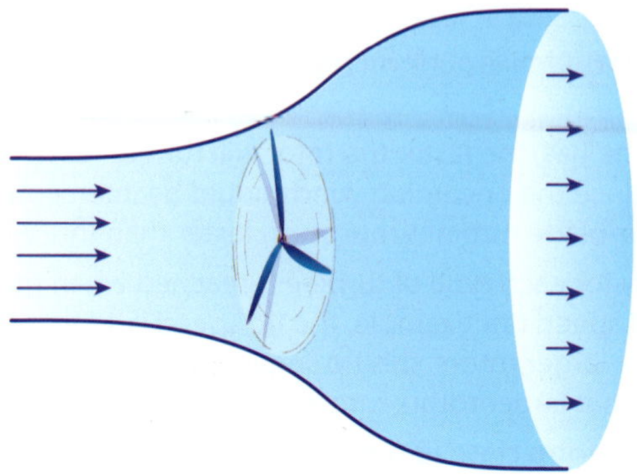

The wind flow profile around a wind turbine. Since the speed is reduced, the profile widens, thereby influencing neighbouring turbines that are too close.

approximately 7 rotor diameters of space between. If we include some extra space around the wind farm – to prevent interference with navigation, for example – a rule of thumb is to have an area of 10 rotor diameters in all directions per wind turbine. This means that the power generated per *horizontal* square metre is about one percent of the wind power per *vertical* square metre. So, if we want to express the power of a wind farm in the number of watts per *horizontal* square metre, the above-mentioned 100 to 200 W/m^2 (a typical value for wind turbines along the coast) falls to 1 to 2 W/m^2, or 1 to 2 MW/km^2, depending on the exact location and size of the wind turbines (for wind farms consisting of large turbines at sea, this may be up to 3 MW/km^2).

This is much smaller than the solar PV cell yield per unit area, but somewhat larger than the yield of plants and trees by means of photosynthesis. Note that wind power is mechanical power and can be converted to electrical power with very little loss.

Obviously, wind farms can be combined with the energy generated by solar cells or photosynthesis. This would attractively allow double use of certain areas on land or sea.

Wind farms

- The world's largest off-shore wind farm began operation in September 2010. It is the Thanet wind farm, located in the North Sea, off the coast of Kent in the UK. It consists of 100 Vesta-90 turbines of 3 MW each, with a (maximum) total generating capacity of 300 MW. They are expected to work at 35% to 40% of maximum capacity (which seems somewhat optimistic). Thus, the total average output of the wind farm can be estimated to be 100 MW.

The turbines are placed in rows 800 m apart, and within each row they are 500 m apart. Since these turbines have a rotor diameter of 90 m, the spacing is a bit tighter than the usual rule of thumb would suggest. If we do not include any extra buffer space around the farm, the expected average power per km^2 is 2.5 MW/km^2. If we do include this extra space for reasons of safety, this figure will decrease to 2 MW/km^2.

- There are two Dutch wind farms located in the North Sea.
The first one began operating in 2007 and is located near Egmond aan Zee, 10 to 18 km off the coast. It consists of 36 turbines of 3 MW each, with a total generating capacity of 108 MW. The average expected power is 30 MW. The farm is comprised of a total area of 27 km^2. This means that it can potentially generate a bit more than 1 MW per km^2.
The second Dutch wind farm, the 'Princess Amalia Windpark', is located 25 km off the coast of IJmuiden. It consists of 60 Vestas V-80 turbines of 2 MW each, for a total of 120 MW. The average output power is expected to be around 50 MW. The rotor span of this type of turbine is 80 m. They are placed at distances of 550 m or 7 rotor diameters from each other. The total wind farm occupies a net area of 14 km^2, but, for safety reasons, another 0.5 km is kept clear along the farm's entire perimeter. This brings the average power per unit area to about 2 MW/km^2.

The conclusion is that wind farms in the North Sea may produce an average of about 2 MW of power per km^2, depending on the location and size.

D | Contribution, Variability and the Cost of Wind Power

Contribution
The deployment of wind power varies greatly by country. In 2009, the total cumulative *installed* wind power was, for example, 25.8 GW in Germany, 19.1 GW in Spain, 4.0 GW in the UK, 3.5 GW in Denmark and 2.2 GW in the Netherlands. The total in the EU-27 was 74.8 GW (data from the European Wind Energy Association, EWEA). In comparison, the installed wind power capacity in the US at the end of 2009 was 35 GW.
Note that these figures correspond to nominal (or 'installed') power.
The average power *delivered* as a yearly average may be a factor of 3 to 5 lower, depending, among others, on the wind conditions in any given year. *The EU* generated 4.2% of total electricity by means of wind power in 2008.

World-wide, the total of all wind turbines installed by the end of 2010 had a maximum generating capacity of almost 200 GW, see figure on p. 124.

These turbines generated a total of 340 TWh in 2009 (data from World Wind Energy Association). The latter figure corresponds to a year-round average of 38.8 GW of power generated. The contribution to total global electricity consumption was 2% in 2009. Note that this is only about 0.7% of *total* energy use.

Variability of wind power
Obviously, wind power suffers from dramatic fluctuations owing to the variations in wind speed. Since wind power is proportional to wind speed cubed, relatively small fluctuations in wind speed can cause quite large fluctuations in the wind power supply, which can be added to the considerable variations in *demand* that can occur within the timeframe of an hour. An example of such variations in demand is shown in chapter 12

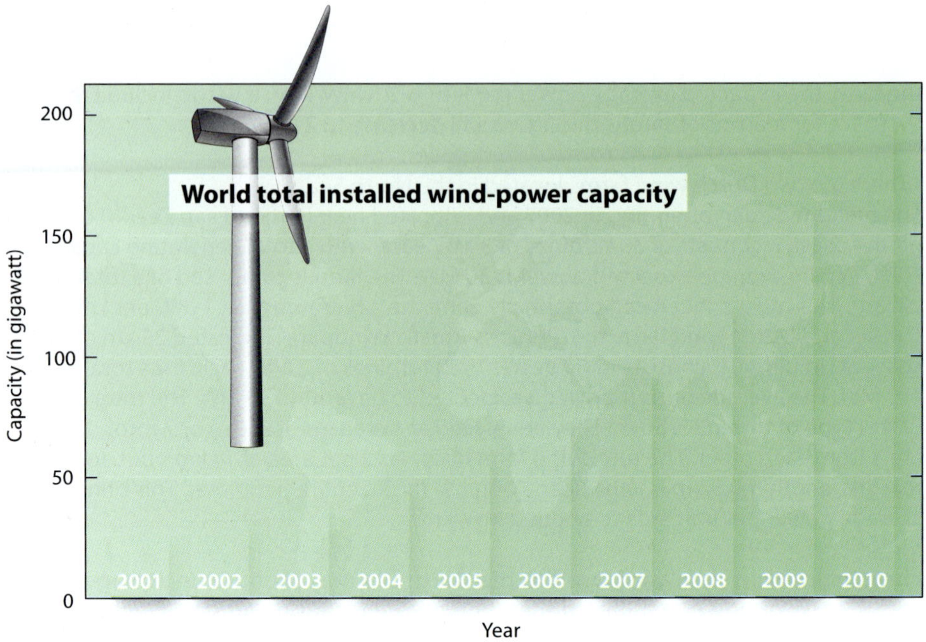

World-wide total installed wind-power capacity for the years 2001 through 2010 (from World Wind Energy Association: www.wwindea.org)

(see the figure on page 167). Power plant operators must continuously balance the total electricity supply with total demand. This requires 'load-following' and rapid-response generators, which have short start-up times. An example of this type of rapid-response facility is a hydropower plant (which has start-up times on the order of a few seconds to a few minutes) or a gas turbine facility (a few minutes for start up). Note that neither base-load power plants like coal plants, which have a start-up time of several hours, or nuclear plants (see chapter 11) qualify as rapid response.

This means, first of all, that the generating capacity of existing plants cannot be reduced when wind turbines are connected to the grid. Second, it means that, to compensate for sudden fluctuations in wind power, flexible (but less efficient) units must be used. This can seriously reduce the 'carbon-saving' effect of wind power generation.

How much wind power can the current grid accommodate? The variability of wind power limits the maximum utilization of wind power when no large-scale storage facility is available. Part of the problem can be solved by the development of large-scale grids and international power exchange. A shortage of wind energy at one place may be compensated by a surplus at another. International exchange also enables storage capabilities to be used more efficiently, for example, by filling a hydropower reservoir in Norway when the wind is blowing in Denmark. Subsequently, Norwegian hydropower can then be used in Denmark at a later time in return. An example of this kind of international cooperation within the EU is the 'NorNed' power cable that runs between Norway and the Netherlands. It can, for example, transport Dutch power surpluses to Norway so that it reduces hydropower generation in Norway, keeping water behind the system's hydroelectric dams for later use. However, one should realise that the energy lost in transmission reduces the overall benefit of international exchange of electricity.

A study conducted by TenneT, Europe's first cross-border grid operator, in cooperation with Delft University of Technology concluded that variations in large-scale wind power in the Netherlands of several GW can be accommodated by conventional plants even without storage systems (B.C. Ummels, Ph.D. thesis, 2009). Closely watching the weather forecasts is a prerequisite.

In the United States in 2008, a study by the Department of Energy concluded that up to 20% of US electricity needs could be met by wind energy by the year 2030 without any advancement in energy-storage technology. This is fairly close to assessments of the situation in the Netherlands. The study notes that regional power generation grids are structured such that there is no need for a full back-up for each individual generator in the system.

Cost of wind power

The energy-payback time of modern wind turbines is about half a year, meaning that after half a year a turbine will have generated the amount of electricity needed for its construction, installation and maintenance. The *financial* pay-back time is much longer: about 7 to 8 years (data from H. Snel, ECN, Petten, the Netherlands).

The cost of electricity production is around € 0.06 to 0.08 per kWh for wind turbines on land and € 0.12 to 0.18 off-shore. (P. Eecen, ECN, 2010). The difference between the two is due to the higher costs for both installation and maintenance of off-shore turbines. These costs are much higher than for electricity generated by coal or gas-fired plants (around € 0.05 / kWh).

In addition to these financial costs, wind turbines cause visual and noise pollution. They can also interfere with fauna such as fish, birds and bats. Bats, in turn, play an important role in the control of mosquitoes and other insects.

New developments: flying windmills

A way to avoid some of the drawbacks of wind power is to take advantage of the high and relatively constant wind speeds at higher altitudes. This may even include the steady jet stream at about 10 km height. Research into this type of power is being performed at several places, *e.g.*, at Delft University of Technology, initiated by former astronaut Wubbo Ockels. Tethered wind kites of 25 and 50 m^2 have been tested which produce over 20 kW of power. Such 'High Altitude Wind Power' or 'Flying Electric Generator' (FEG) technology may be able to produce electricity at low cost. Clearly, the potential of these flying windmills is huge, but its technical and economical feasibility has yet to be demonstrated.

Conclusions regarding Wind Power

- Since wind power is proportional to the wind speed cubed, wind turbines should be placed at especially windy locations: along the shore or at sea.
- Modern wind turbines achieve almost maximum possible efficiency, so there is not much room for improvement.
- As a rule of thumb, wind turbines along the shore produce 100 watts per m^2 of the area swept by the rotor. For large off-shore turbines, this is 200 W/m^2.
- Since wind turbines may be in each other's lee, they need sufficient room. For a row of turbines oriented toward prevailing winds, a distance of 2 to 3 rotor diameters between turbines suffices. For wind farms spread across a grid, more space is needed. This results in an annual average yield of 2 MW per km^2 for wind farms located in the North Sea.
- Replacing a conventional 1,000 MW plant in continuous operation requires about 1,000 large wind turbines with a nominal power of 3 MW, having a rotor diameter of 80 to 90 metres.
- Wind turbines need about half a year to generate the amount of electrical energy needed to recover the energy used in manufacturing and installation. However, the *financial* pay-back time is much longer at 7 to 8 years. The life expectancy of a turbine is 15 years.
- The cost of wind power is around 0.06 to 0.08 €/kWh for wind turbines on-shore and 0.12 to 0.18 €/kWh off-shore.
- Fluctuations in wind power can be partially accommodated by using large-scale (international) electricity grids. A surplus in wind energy can then be used to reduce the output of hydroelectric dams, which can store water to be used later. Storage seems unnecessary as long as wind produces less than 20% of total electricity needs.
- Fluctuations in wind power require rapid-response generators as back-up units. This will partially negate the fossil fuel and CO_2 emissions savings.
- Other drawbacks of wind power are noise, and the effects they have on fauna such as birds, bats and fish.

10 | Energy from Water and from Soil

The major portion of the renewable energy fluxes from water and soil are indirectly generated by the sun (just like wind power). This means that, on a global level, they are much less significant than direct solar power derived from the sun's radiation. Nevertheless, they are important because they deliver solar energy in a more concentrated way. This is obvious in the case of hydropower because it uses the potential energy of water that evaporates from the ocean over a large area, and is eventually collected by streams, rivers and lakes.

A | Hydropower

Hydropower began with the development of water mills, long ago. However, these days hydropower is generated by vast artificial lakes and gigantic turbines, having developed into the world's most important renewable energy source by far. The largest hydropower suppliers are China, Canada, Brazil, the US and Russia. These five countries produce more than half of the world's total hydropower.

Hydropower's contribution to global electricity production in 2007 was 16% (3.16 out of 19.85×10^6 GWh; source: International Energy Agency, 2010). This value corresponds to 360 GW on an average continuous basis, or the equivalent of 360 large thermal power plants in continuous operation. This is even slightly more than the total nuclear power generated world-wide.

According to research by the World Energy Council, only about 1/3 of the hydropower potential is currently being utilised. Opportunities for additional growth are primarily located in Africa (an increase by a factor of 10), Asia (by a factor of 3), South America (by a factor of 2) and North America (by an additional 10%).

It is, however, doubtful whether hydropower's full potential will indeed ever be utilised. We need to be aware that the building of large-scale dams can have a dramatic impact on the environment, since huge areas have to be first vacated and then flooded.
An example is the *Three Gorges Dam* in the Yangtze River in China, the world's largest hydropower project. Its electricity-generating capacity after completion in 2011 is huge: 22,500 MW, which is about the total electricity-generating capacity of a country like the Netherlands. However, its reservoir with a surface of more than 600 km^2 flooded archeological and cultural sites and displaced over 1.3 million people. As a consequence, the dam has been a controversial topic both in China and abroad.

In the EU hydropower produces roughly 10% of the total electricity generated. Hydropower is primarily generated in Norway, France, Italy, Austria, Spain, Sweden and Switzerland. This list may not come as a surprise: these are all countries with high mountains.

Options within the EU for further development of hydropower are fairly small, however. In relatively flat countries like Denmark or the Netherlands the use of hydropower is obviously very limited, despite the many rivers that may flow through such countries. One should bear in mind that a mere flow of water as such is not sufficient to produce energy. A certain amount of pressure is also necessary, which requires height differences and dams.

In *developing countries*, where energy use is still relatively low, hydropower can be an important energy source. An example is Egypt which uses the Aswan Dam in the Nile, about 1,000 km south of Cairo. It was originally designed in 1902 mainly to regulate the river, to prevent flooding and to store water for agriculture. Presently the two Aswan dams have a total installed capacity of 2.1 GW.

Advantages and drawbacks of hydropower
Hydropower has obvious advantages. It is cheap (otherwise it would not have seemed so attractive during the eras when oil was still relatively cheap). Hydropower is also clean, producing no CO_2. Moreover, it has a short start-up time, and provides its own storage if the hydroelectric plant has its own dam. It can, therefore, easily adjust to electricity demands, in contrast to wind and solar power. Therefore, hydropower on the one hand, and solar and wind energy on the other can be considered to be complementary energy sources.

Hydropower's *drawbacks* include the large impact that building a dam and a hydroelectric plant have on the local environment and local population (see, for example, the *Three Gorges Dam* in China on p. 127). Furthermore, there is the risk of dams breaking as a result of an earthquake, possibly affecting the existence of countless people living downstream. Finally, hydropower can also cause a temporary but substantial contribution to global warming by the decay of vegetation under water, which produces large quantities of methane (CH_4).

B | Wave Energy

Wave energy is basically concentrated wind power. If wind sweeps across a large body of water, waves can build up and store large amounts of energy. This is because the energy stored in waves of water does not readily dissipate in deep water. Furthermore, bear in mind that the mass density of water is a thousand times greater than the density of air.

The most practical application seems to be the harvesting of wave energy along the coast. Therefore, coastal regions with prevailing wind heading towards the shore can serve as prime locations. Since waves need large distances to accumulate their energy, one would expect for example the UK's west coast to be well-suited for the development of wave energy.
A world map showing wave power potential for different regions reveals that this is indeed the case (see figure on page 129). For instance, along the Norwegian, British and French coasts, this potential is estimated to be on the order of 50 kW per metre of coast.

At first sight this may seem a surprisingly high number. But by using a bit of high school physics we can easily conclude that the power of waves is potentially large indeed. Let us look at one metre of coast, and let us assume the waves are about 3 m high. If we estimate the volume of water in such a wave to be 10 m^3, the potential energy is about 300 kJ/m of coast. If such a wave were to pass every 6 seconds, we would end up generating about 50 kJ of power per second, or 50 kW. Note that this is the power of a medium-sized car.

If we were able to convert this power into electricity at an efficiency of 50%, we would be able to replace a large 1,000 MW power plant by using 40 km of coast, which is, however, still a large stretch of coast.

As seen from this world map, the coasts of Western Europe do reasonably well, compared to many other places around the globe.

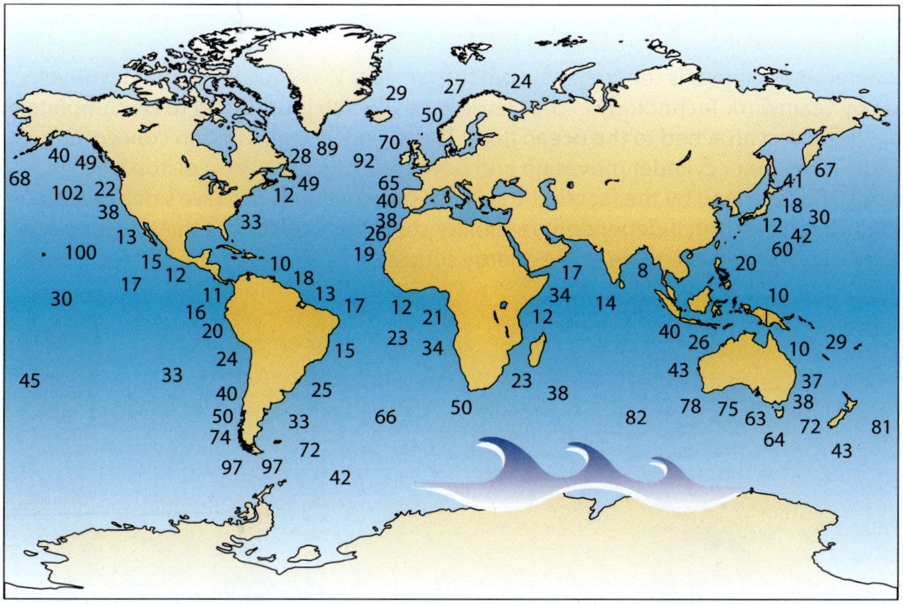

Wave energy (in kW per metre)

Average wave energy potential in terms of kilowatts per metre of coast for different locations around the globe (from: World Energy Council)

The economically recoverable wave energy is estimated to be 2,000 TWh/year worldwide, or an average continuous power of about 200 GW, or the equivalent of 200 large power stations. Although this is potentially a lot of power, it would still be less than the amount generated by hydropower plants.

The key question is whether wave energy can be harvested at a reasonable level of efficiency. Although this might be possible, it is not easy. There are a few aspects which make the harnessing of wave energy difficult.

- The equipment needs to be designed so that it can achieve a maximum efficiency at typical power levels for the location (50 kW/m, for example) and yet be able to survive a heavy storm.
- The direction of the rolling waves varies according to the direction of the wind. This makes things complicated on the open sea. This is less of a problem closer to the shore, since waves tend to mimic the coastline.
- One problem that it shares with all other movable, mechanical devices at sea is corrosion.

To date, a few pilot plants that follow their own operating principles have been built. Some plants like the *PowerBuoy* (see photo), for example, employ floating elements which float on the moving waves. The up-and-down motion of the buoy drives a generator whose electric power is transported back to shore via underwater cables. According to the manufacturer, a 'field' of these buoys spread across 1/8 km^2 (30 acres) can generate 10 MW of power.

A comparable concept is the 'water swing' (*Archimedes Wave Swing*), a product developed by Teamwork Technology in the Netherlands, which is designed to be completely submerged and attached to the ocean floor. The 'swing' consists of two concentric cylinders. The inner cylinder moves up and down on the waves like a piston (see illustration). This is caused by the fact that the water pressure under a wave's peak is greater than under a valley. An independent feasibility study concluded that this may become an economically feasible sustainable energy source.

The PowerBuoy by Ocean Power Technologies (Source: www.oceanpowertechnologies.com)

Archimedes Wave Swing (Source: www.reuk.co.uk)

Another pilot project is the *'Tapered Channel Device'*, or *'Tapchan'*, which employs tapered channels that force the incoming water wave upward into a higher reservoir, using the water's own momentum. The water subsequently flows back to drive a turbine (see illustration). One advantage of this project is that it employs existing salt-water turbine technology.

Tapered Channel Device utilised to harvest wave energy
Its shape forces an incoming flow of water upward.
Energy is then generated by the backflow.
(Source: http://re.emsd.gov.hk)

Another approach that is *not* affected by salt-water corrosion is the *'Oscillating Water Column'* (see illustration below). This device converts rising and falling water levels into the compression and expansion of air, which, in turn, drives a 'Wells turbine'. This type of turbine has a fixed sense of rotation that is independent of the flow direction of the air that drives it. A small power plant called the *Islay Wave Power Station*, which employs this principle was built on the isle of Islay, off the Scottish coast. It generates 0.5 MW of electricity. None of these pilot projects are as yet considered economically feasible.

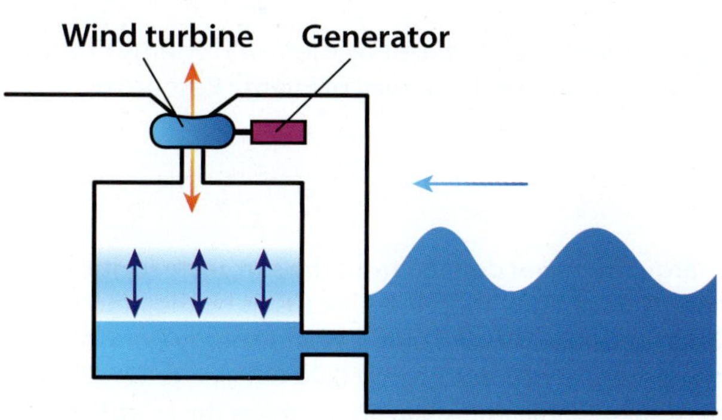

The Oscillating Water Column may be a possible device for harvesting wave energy. The waves force the water level (left) up and down, which in turn pushes and pulls air through a wind turbine that drives a generator.

The Aguçadoura Wave Park off the Portuguese coast (Source: www.pelamiswave.com)

The *Aguçadoura Wave Park* is a commercial wave power plant that has been in operation off the northwestern coast of Portugal since 2007. It currently generates over 2 MW and consists of a number of coupled, partially submerged cylinders which move up and down on the waves, called the *Pelamis wave energy converter.* The motion of the cylinders relative to one another is relayed to an electricity generator via pivoting joints and hydraulic engines. The plant is designed to withstand even heavy storms, which, however, somewhat compromises its efficiency, basically preventing it from generating the maximum amount of energy from the movements of the waves.

A survey of various wave energy systems was produced by the World Energy Council and can be found at www.worldenergy.org (under Survey of Energy Resources).

C | Tidal Energy

This is a form of sustainable energy that is *not* derived from the sun. Instead, it is better characterised as 'moon energy', since tides are primarily caused by the moon. The global potential, however, seems miniscule: 5 GW (Source: World Energy Council). This is much smaller indeed than the potential of wave energy (200 GW, see p. 129). Nevertheless, it is an appealing form of sustainable energy in areas where the coastline has a funnel-like shape. This forces incoming water up to an extra high tide, thus creating a natural concentration of tidal energy.

An example is the *Barrage de la Rance*, near St. Malo in France, at the mouth of the river *Rance*, where the average tidal difference is 8.40 m. A tidal power plant was built there in 1966, which generates 240 MW of power. The accompanying tidal lake has a surface area of 22 km^2.
Another location well-suited to tidal energy development is the funnel-shaped mouth of the Severn, near Bristol, UK, where the tidal differential is up to 10 m. Plans to build a tidal power plant have been under discussion for quite a while (since 1849, to be precise).

There is an interesting feature inherent to tidal power plants. It is profitable to pump extra water into the tidal lake when the tide is highest. The reason is the following: suppose the tidal differential is 9 m, and we raise the level of the tidal lake by one metre to 10 m at its highest point. Each m^3 of water that has been raised by 1 m will later – at low tide – fall the full 10 m and thus deliver net energy.

D | Temperature Differences in the Oceans (OTEC)

The top layer of the ocean is always much warmer than the water deeper down. This temperature differential may be utilised by designing a device such that the two thermal reservoirs drive a thermal engine that produces mechanical energy. This is the Ocean Thermal Energy Conversion (OTEC) system. The temperature difference is relatively small, so the efficiency of the cycle is bound to also be small – 3% maximum (see chapter 2). A positive aspect is that it has a very large capacity.

Research into this form of sustainable energy began in 1979 on Hawaii (along Big Island's Kona coast). An OTEC plant was built there in 1999 with a maximum of 250 kW, or the combined power of five medium-sized car engines. This maximum is reached in late summer when the water's surface is the warmest. The cold water is then pumped from a depth of 800 m through pipes 1 m in diameter. Plans to build a larger 1-MW plant have thus far not been enacted. Other countries, such as Japan, have plans for large OTEC plants. Nevertheless, this remains a relatively unimportant concept.

E | Osmosis: Salt and fresh water ('Blue Energy')

Whenever fresh water and salt water are separated by an ion-selective membrane, one ion will pass through the membrane more easily than the other. In the case of ocean water, where the salt is mainly NaCl (sodium chloride), the ions in question are positive Na ions and negative Cl ions. Such a selective transport process occurs continuously in our own bodies. And just like in our bodies, a modest potential difference – about 80 mV – builds up across the membrane. But by combining many of these cells into a larger unit, a higher potential can be generated. The principle is the reverse of dialysis, and is, therefore, sometimes called *Reverse Electro Dialysis*. It creates opportunities for generating energy in places where a (fresh-water) river flows into a (salty) sea. The principle is sometimes referred to as 'Blue Energy'.

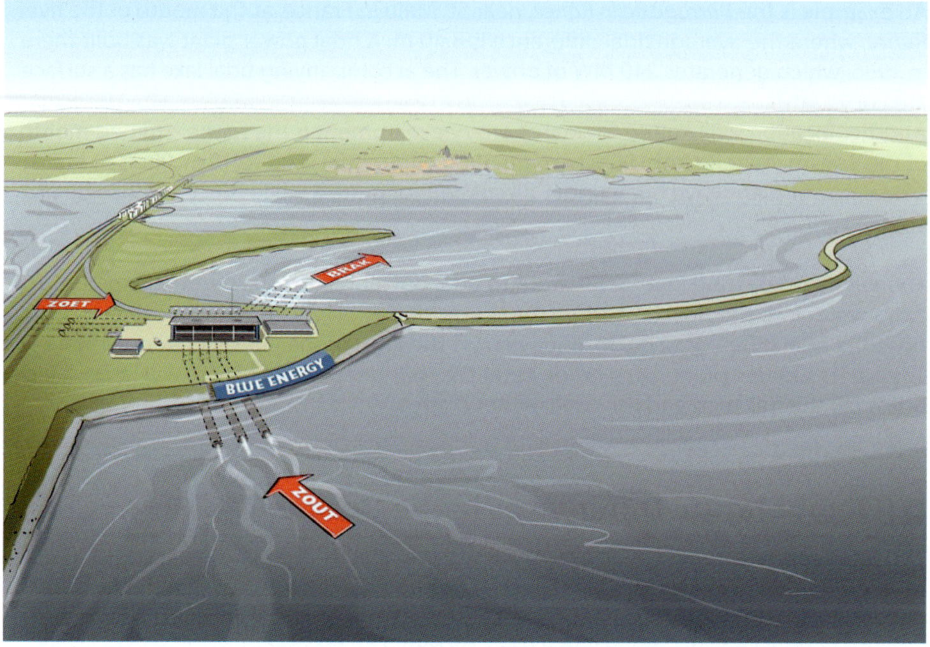

Artist's impression of 'Blue Energy' using the principle of osmosis (Source: Rijkswaterstaat, the Netherlands).

This may be an interesting option, for example where the Rhine empties into the North Sea. The corresponding energy is comparable to the energy generated by a hydro-power plant from the Rhine's waters flowing into the sea at a height of 240 m. However, a National Academy of Sciences (KNAW) study predicts a contribution to the Dutch electricity production of a few percent at best.

An alternative location is the main IJsselmeer dike ('Afsluitdijk'), which essentially separates fresh water from salt (see artist's impression above).

However, it is not yet clear whether this idea will ever be technically and economically feasible on a large scale. The system must still overcome some technological challenges such as the production of cheap membranes, keeping them from getting clogged in dirty water, decreasing their resistance and increasing their life expectancy.
The 'Wetsus' institute for water technology is currently working on this project (see www.wetsus.nl).

Besides this 'electrical' version of osmosis, there is also one based on differences in pressure, which arise when fresh water passes through membranes more easily than salt water. One can use this pressure difference to drive a turbine, which, in turn, powers a generator. The large-scale feasibility of this version remains unclear.

F | Geothermal energy

The centre of the earth is pretty hot: 4,000°C. So there is a permanent flow of heat from the earth's core towards its surface. This heat is primarily owing to radioactive decay of thorium, uranium and potassium isotopes which have long half-lives. There is also some heat flow which is a remnant of the earth's hotter early days. The total heat flux is small, averaging 0.06 watts per square metre.

This can be checked on the back of an envelope: at a depth of 1 km, the temperature is almost everywhere about 30°C higher than at the earth's surface (even higher temperatures would mean miners would be unable to survive at these depths!). This means that there is a temperature gradient of about 0.03 degree per metre. If we multiply this by the thermal conductivity coefficient of the earth's crust (circa 2 $Wm^{-1}K^{-1}$) we end up with a heat flux density of 0.06 W/m^2.

This geothermal heat flow to the earth's surface is very small compared to the solar heat flux, which is roughly 200 W/m^2 on average (see figure on page 95). This is about 3,000 times greater than the geothermal heat flow.

However, this small sustainable geothermal heat *flow* to the surface of the earth is accompanied by large *quantities* of heat that have accumulated at various locations inside the earth over time. This geothermal heat supply may be significant, especially in relatively young volcanic regions. Hot rock is found at depths of 1 to 3 km – sometimes accompanied by hot water or steam – at temperatures of 200°C or more. At such high temperatures it is commercially viable to generate electricity using 'geothermal power plants'. However, this is not economically feasible at temperatures below 150°C, since the conversion efficiency is too low (see chapter 2). These lower temperatures are, however, well-suited for heating purposes. Thus, it is useful to distinguish two types of geothermal energy derived from (exhaustible) geothermal heat supplies.

High-temperature geothermal energy
It is profitable to build geothermal power plants with temperatures over 150 to 200°C. Some examples:
Italy has produced electricity from geothermal steam since 1904, in Larderello, Tuscany. The first 'commercial' geothermal power plant began operation in 1913. As of 2009, the Larderello geothermal field had 21 operating units with an installed capacity of 560 MW.
Iceland's geothermal energy is being used for both electricity generation and for heating purposes. The Nesjavellir Geothermal Power Station supplies Reykjavik with most of its hot water (1,800 litres per second!) and also generates 120 MW of electricity.
The United States has the largest complex of geothermal power plants in the world, located at The Geysers, about 100 km north of San Francisco. It consists of 18 geothermal power plants, having an operating capacity of over 800 MW and generating about 7,000,000 MWh of net electric power per year on average. It forms part of the Mayacmas mountains geothermal system, east of the San Andreas Fault.

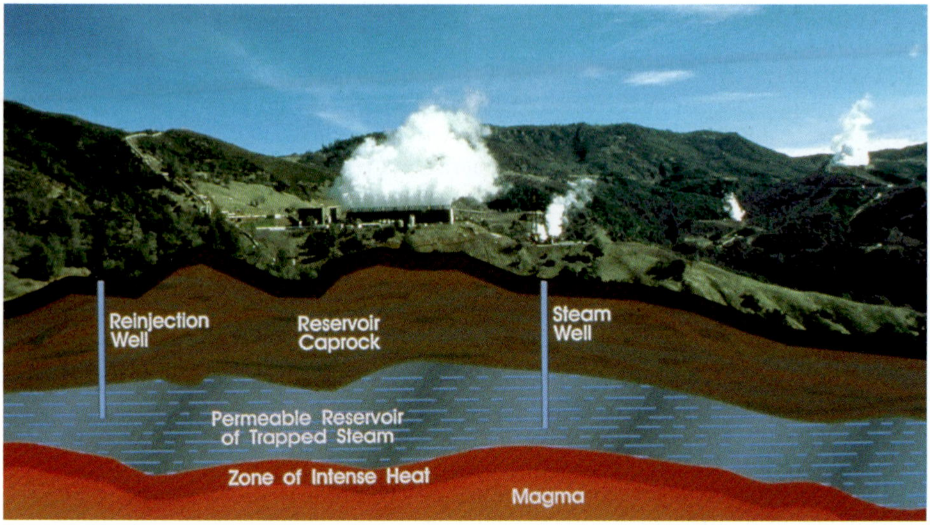

The Geysers in California, USA. This figure illustrates the principle of generating electricity by injecting water into hot fractured rock, where it is converted into pressurised steam, which drives a turbine. (Courtesy of Calpine Corporation: www.geysers.com)

Natural geothermal steam was produced for many years and steam condensate was used for evaporative power plant cooling. However, rainfall couldn't seep into the reservoir fast enough to keep the deep underground steam reservoir charged with pressure and steam flow. Seasonal creek water runoff and treated waste water from local cities are now injected to help replenish the steam reservoir and sustain generation. Water is injected into the hot, permeable rock about 2–3 km deep, where heat from even lower-lying magma converts it into pressurised steam. The steam is tapped by steam production wells and piped overland to a power plant where it drives a turbine-generator to generate electricity (see the image above).

The Geysers permeable reservoir is composed of fractured metamorphosed sandstone (aka 'graywacke') and is capped by an impermeable rock formation. One drawback is that extracting heat from a subterranean hot reservoir causes it to contract, which creates micro-earthquakes, or so-called *Induced Seismicity.* (Source: J. Farison, Calpine Corporation.)

Electrical generation from high-temperature geothermal reservoirs is only part of geothermal power's potential in the US. The 2006 MIT report *The Future of Geothermal Energy* concluded that the US can potentially generate about 100 GW of geothermal electricity economically, or the equivalent of 100 large conventional power plants.

In addition to geological formations that have natural fracture permeability, non-permeable heat reservoirs are also currently being exploited by Enhanced (or: Engineered) Geothermal Systems (EGS), whose technology involves the creation of a network of fractures in hot rocks, through which water can be pumped.

This technique would not limit the generation of geothermal energy to volcanic regions but potentially could be expanded to include large areas of the US. This is undoubtedly also true for numerous other countries.

World-wide, installed geothermal facilities currently generate in excess of 10 GW of electricity. The proportion of geothermal electricity generated in 2007 was 0.3% of the total electricity demand (which is 0.1% of total global energy demand only).
But the potential is enormous: estimates are between 65 and 138 GW of generated electricity. This seems to be a low estimate, given that the US projects generation of 100 GW itself, as noted above. Moreover, there is also the use of geothermal energy for heat, see below.

Several pilot projects are currently being carried out in France (Soultz, Le Mayet de Montagne), the UK (Rosemanowes), Germany (Falkenberg, Horstberg, Bad Urach) and Switzerland (Basel, Geneva).

The nice thing about these new geothermal energy developments is that deep-drill technology can be borrowed from the oil industry. This means that geological formations as deep as 5 km will soon be within reach. This also means that application of geothermal energy may become much more widespread than foreseen only a few years ago.

Low-temperature geothermal energy
Most countries have no active volcanoes, and thus lack the accompanying high-temperature reservoirs. Geothermal *electricity* generation is therefore not an attractive option because depths of 5 km would need to be drilled to reach the desired 150°C temperatures. Nevertheless, using geothermal energy may still be an option if we settle for somewhat lower temperatures. Temperatures well below 100°C are perfectly suited for home-heating needs, especially if floor-heating systems are used.

Given the above-mentioned average temperature gradient of 30°C per km, it is not surprising that temperatures at –2,000 m are between 70 and 90°C in many places. This is more than sufficient for home heating needs. Even in a flat country like the Netherlands, several projects are under way to develop geothermal sources for heating. In The Hague's southwest, for example, about 4,000 homes and several office buildings are earmarked for the use of geothermal heat drawn from –2,200 m, where the temperature is around 75°C. Tests performed in 2010 show that 150 m^3 of hot water per hour can be produced (see, for example, www.geothermie.nl).

Even relatively cold geothermal water can be used as a heat source if heat pumps are used to achieve higher temperatures. The shallower aquifers (natural water reservoirs) are well suited for this purpose. These layers can also be used as a long-term storage system (see chapter 12).

Geothermal energy represents a promising alternative energy source. Its potential is very large and its exploitation does not require large areas of land, since most of the activity occurs underground, while the power plants above ground can remain relatively small. Moreover, much of the drilling technology can be borrowed from the oil industry.

However, as was pointed out above, most geothermal energy is not truly sustainable because it uses heat that has accumulated over many years, and is thus exhaustible. The geothermal heat that is drawn from geological layers takes about 1,000 years to be replenished. Thus, geothermal wells will eventually have to be moved to new locations. Therefore, geothermal energy should probably be considered a semi-sustainable energy source.

Conclusions regarding Energy from Water and from Soil

- Hydropower is very important on a global scale, providing about 16% of total global electricity (or 360 GW on a continuous basis). There are ample possibilities for further expansion of hydropower, especially in Africa and Asia. The large environmental impact and the risk of dams breaking and causing flooding (*e.g.*, by earthquakes) may limit its expansion.
- Wave power at 200 GW world-wide has a lower exploitation potential than hydropower. Moreover, it is still in its infancy.
- Tidal energy has a much smaller potential than wave power (5 GW). It is interesting primarily in places where a funnel-like coastline enhances tidal differences.
- Osmosis between fresh and salt water (known as 'Reverse Electro Dialysis' or 'Blue Energy') may generate some modest amounts of energy. Technical problems concerning membrane technology have yet to be solved.
- Geothermal energy has enormous potential. It is not truly sustainable, however, since it uses geothermal heat that has been stored over the ages. Regions (mostly volcanic) that house high-temperature reservoirs or hot dry rock formations can be exploited for electricity generation. Most regions around the world can utilise geothermal sources for heating purposes, possibly in combination with heat pumps. Drilling technology can be borrowed from the oil industry.

11 | Nuclear Energy

Nuclear fission and nuclear fusion both produce energy. Isn't that odd? First we split atoms (or, more precisely, atomic nuclei), thereby producing energy. Then we fuse the atoms, again producing energy. And so on. Did we finally invent the perpetual motion machine?

This is nonsense, of course. We are talking about completely different kinds of atomic nuclei. The secret is in the binding energy between different nuclei. This is the energy that keeps the nuclei from falling apart. The binding energy differs for different nuclei. This is nicely illustrated if we look at the *mass* for different atoms, which reflects the binding energy. It turns out that there is something funny about the mass. The building blocks of atoms (protons and neutrons) have different masses when they are embedded in different nuclei. We can see that clearly if we look at the mass of atoms divided by the number of building blocks (or 'nucleons', *i.e.*, protons and neutrons) that they contain. We might expect to get the same value, but what we get is something different. This is illustrated in the figure below, where this mass per building block is shown as a function of mass number, which is the total number of protons plus neutrons in a nucleus.

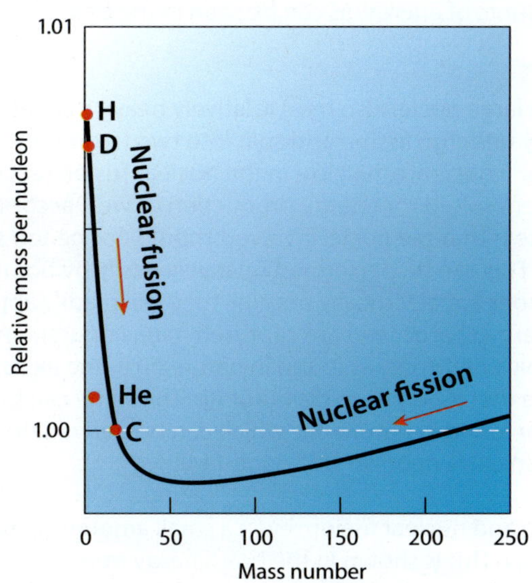

The mass of different atomic nuclei divided by the number of building blocks (protons and neutrons), relative to the value for the carbon-12 nucleus (which contains 6 protons and 6 neutrons).

This graph shows that ordinary carbon – which contains 12 building blocks: 6 protons and 6 neutrons – has an atomic mass number of 12. So, carbon has a 'mass' of 1 per building block. (The actual mass is a very small fraction of a kilogram, namely 1.67×10^{-27} kg, but we will use it as the basic unit, u, on the atomic scale.)

We see that other atomic nuclei are, somewhat surprisingly, just slightly different. Very small nuclei, like normal hydrogen (H) or 'heavy hydrogen' (D, for deuterium) turn out to be a bit too heavy. The difference is very small – not even 1% – but the building blocks of these nuclei are noticeably heavier than, for example, helium (He) or carbon (C). Average-sized nuclei, like iron, are the lightest, while the mass per building block increases again for larger nuclei like uranium.

This shows that there are, indeed, two types of nuclear reactions that yield energy.

- Nuclear fusion. Consider deuterium, D, which consists of 1 proton and 1 neutron and has a relatively large mass per building block. Now look at helium (He), which consists of 2 protons and 2 neutrons, so it is essentially two D-nuclei put together. However, it is lighter than the two D-nuclei. So, if we were to fuse two Ds to form one He, some small bit of the mass would disappear. This difference in mass is converted into energy, according to Einstein's famous formula $E = mc^2$, where E is energy, m is mass and c is the speed of light. If we plug in the numbers, the mass difference is 4.0282 – 4.0026 = 0.0256 in the above-noted atomic mass units. This is an extremely small amount of mass. However, since c is such a large speed (not to mention c^2 !) these tiny masses correspond to large amounts of energy, as can be seen in the box 'Energy from nuclear reactions'.

- Nuclear fission. Very large nuclei also have relatively heavy masses, as seen in the figure on p.139. If they undergo fission and split into two fragments of roughly equal mass, the *total* mass will decrease, just like in the nuclear fusion reaction of very light nuclei. This makes them suited for energy production as well, according to Einstein's $E = mc^2$. A prerequisite is that the nuclei involved should not be too stable. After all, they have to be split. This can be performed relatively easily by bombarding them with neutrons. Since neutrons are electrically neutral, they will readily approach these heavy nuclei, despite their large charge, and are captured here. For uranium-235 (which has a total of 235 protons plus neutrons, as found in nature) just one extra neutron is enough to make it so unstable that it begins to disintegrate. The same can be done with plutonium-239, which is *not* found in nature but which can be derived from uranium-238 if it manages to capture an extra neutron (see page 144).

In both nuclear fusion and nuclear fission, even a small amount of mass can generate a large amount of energy. This is shown in the box 'Energy from nuclear reactions' where a mass of 0.001 kg that is converted into energy is found enough to heat 200,000 m^3 of water from 0°C to 100°C. This large amount of water would fill countless bathtubs; it would, in fact, fill a large football stadium with hot water.

Energy from nuclear reactions

Just how much energy is released in nuclear reactions becomes clear if we do the calculations. Suppose that 0.1% of the mass involved is converted into energy (and this is not an unreasonable assumption, as seen in the figure on p. 139) and let's take 1 kg of nuclear fuel. This means that 0.001 kg will be converted into energy. According to Einstein's $E = mc^2$, the energy released is equal to

$E = 0.001 \times (3 \times 10^8)^2$ joules $= 9 \times 10^{13}$ J.

This is not a very practical number, however. If we express it in terms of oil (having a heat of combustion per litre of about 40 MJ or 4×10^7 J), we see that the energy released by converting a single gram into energy is the equivalent of about 2 million litres of oil. Suppose we were to use this amount of energy to heat water from 0°C to its boiling point, how much water would that be? Since heating 1 m^3 of water by 100°C takes 4.2×10^8 J, we find that the above energy permits us to heat over 200,000 m^3 of water to its boiling point, which is the volume of a large football stadium.

A | Fission

The current nuclear technology employed world-wide is based on nuclear fission. All 440 of the world's nuclear reactors use this principle to generate electricity. More precisely, the reactors do not produce electricity, but heat. The heat is used to drive a steam turbine that, in turn, powers a large generator. So, the specific nuclear aspect is at the initial step: generating heat through nuclear reactions.

Several types of fission reactors are currently in use, but it is useful to distinguish the two major approaches: 'ordinary' nuclear reactors and breeder reactors. Almost all current reactors in operation are 'ordinary' reactors. They use primarily uranium-235 as a fuel, which is the less abundant uranium isotope (for details, see below). The idea of *breeder* reactors is to also use the much more abundant uranium-238 isotope, which makes up 99.3% of natural uranium. To be fissile it must be converted – in a few intermediate steps – into plutonium-239, which is fissile like uranium-235 (see the section Breeder Reactors on p. 150).

A1 | Current Fission Reactor Technology

Natural uranium consists of 0.7% of the fissile isotope U-235. The other 99.3% is U-238 (the numbers 235 and 238 refer to the total number of 'building blocks', *i.e.*, protons plus neutrons, that make up the atomic nucleus). The difference between the two is the number of neutrons. Both have 92 protons in their nuclei. Since the total atom must be neutral, both also have 92 electrons spinning around the nucleus. And since the electron shell determines the chemical properties, the two are chemically identical. So, we find them at the same spot in the periodic table of the elements, which is why they are called 'isotopes' (from Greek *iso* = same and *topos* = place).

The fact that both isotopes of uranium are found in nature is a pretty obvious sign that they have long lives. The half-life of U-235, for instance, is 0.7 billion (0.7×10^9) years, and for U-238 it is even longer: 4.5 billion years. (The concepts of 'half-life' and radioactivity are discussed in more detail on page 162.) It is, by the way, fairly easy to comprehend why the isotope that decays the slowest is also the most abundant – U-238 – if we assume that the two started off equally abundant when the earth was formed.

It is precisely because of their slow decay that both uranium isotopes emit only very low levels of radiation. The reason is that the radiation emitted in the decay process is spread out over billions of years. This makes it possible for uranium to be mined in uranium mines without serious risks to the miners, provided that they keep uranium outside their body. Uranium emits α-particles and is hazardous when inhaled or ingested (see section Radioactivity and Health Hazards, page 159).

The fissile isotope is U-235, and if the nucleus of a U-235 atom gets hit by a neutron, it may disintegrate. Generally it splits into two mid-sized fragments plus an average of 2.5 neutrons:

$$\text{U-235} + \text{n} \rightarrow \text{X} + \text{Y} + (2 \text{ to } 3)\ \text{n} + \text{Energy}.$$

Those 2 to 3 neutrons can, in turn, hit and split other nuclei, which produces new neutrons, etc. This creates the possibility of a chain reaction.
This is schematically illustrated in the following figure. In this case the related fragments are one barium and one krypton nucleus, but there may also be different fragments.

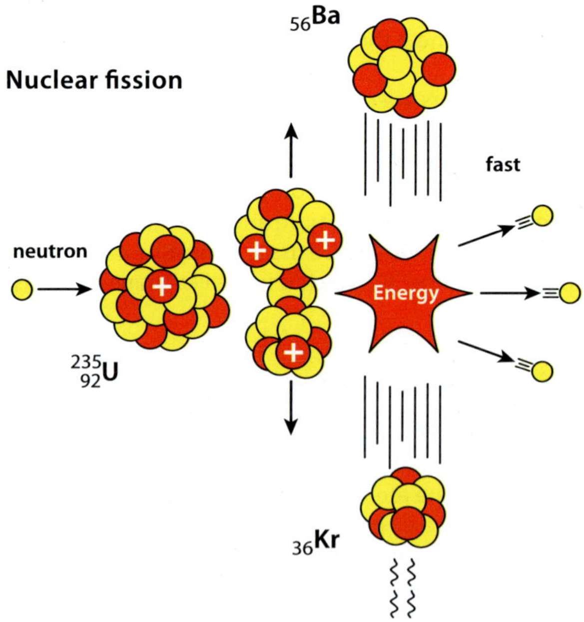

The principle of a nuclear fission reaction. A uranium nucleus (U-235) is hit by a neutron, splitting it into two fragments plus a few neutrons, which, in turn, can split more nuclei.

The chance that a uranium nucleus will be split by a neutron is great only if the neutron is relatively slow. However, neutrons emerge from the fission process at very high speeds, so they must be slowed down to increase their efficacy. This happens most effectively by means of collisions with particles of the same mass. Just think of two billiard balls. When they collide, they can exchange speeds very readily. Indeed, if a stationary ball is hit by a rolling ball head-on, the stationary ball will absorb and translate the speed of the rolling ball. If the masses are different, this exchange of speeds would only be partially effective. Just think of a pea hitting a billiard ball: it will bounce back at almost its initial speed, while the billiard ball will hardly move at all.
When it comes to neutrons we are lucky, however. Ordinary water (H_2O) has two H-nuclei, which have essentially the same mass as a neutron. It is, therefore, an ideal 'cooling liquid' to handle the neutrons' immense kinetic energy and slow them down. This has two effects: the kinetic energy of the neutrons is transferred to the water (which heats up) and the neutrons are slowed down (in preparation for splitting another nucleus). So the water performs two tasks simultaneously. Most nuclear reactors are 'light-water reactors' (some 'heavy-water reactors', using D_2O, also exist). Moreover, they are often called 'thermal' because the neutrons are slowed down to ordinary, thermal energy, and have speeds that correspond to particles moving around at moderate temperatures (*i.e.*, 300°C).

An example of a nuclear reactor is shown in the figure below. The water that slows down the neutrons and removes the fission energy circulates in the 'primary cooling circuit' (yellow-red in the figure). Since the water temperature is about 300°C, the corresponding vapour pressure is high at 150 bar and that is why these reactors are called Pressurised Water Reactors (PWR). The primary loop exchanges its heat with a secondary circuit (blue) where steam is produced, which drives a turbine and a generator to produce electricity. The steam is condensed by cooling water from outside, which forms a third circuit.

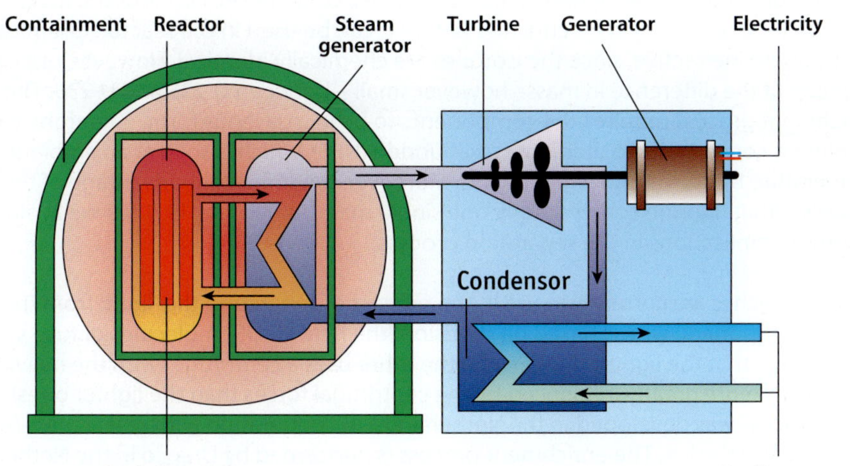

Schematics of a Pressurised Water Reactor or PWR
(courtesy Nuclear Research and consultancy Group, www.nrg-nl.com).

But there is a complication: neutrons released in a fission reaction may get lost by being captured by U-238 nuclei (which are by far in the majority). Even though the chances that this will happen are fairly small for these slow neutrons, it is still too large of a risk to use natural uranium. Therefore, in order to be used in most current reactors, natural uranium is enriched so that it contains 3 to 4% of U-235. This is usually done in ultracentrifuges (see *Fuel Cycle and Possible Hazards* below).
Even if this type of enriched uranium is used, some portion of the neutrons will still be captured by (non-fissile) U-238. If this happens, the nucleus decays, via a few intermediate steps, to a (fissile) plutonium-239 nucleus. Part of this newly formed plutonium (one-third to one-half) will immediately cause fission in the reactor, and thus contribute to energy production. The remainder will eventually exit the reactor with the rest of the nuclear waste.

A2 | Fuel Cycle and Possible Hazards

The complete fuel cycle consists of numerous steps, from mining and enrichment to waste disposal. We will look at three steps in more detail in view of possible risks. These are, first, the enrichment process, since this offers the possibility of developing weapons-grade uranium. Secondly, we look at the reactor itself since it involves the production of highly radioactive material and the risk of serious accidents. Finally, we focus on the radioactive waste, since it needs to be handled and disposed of carefully, and the plutonium content must be guarded to prevent its misuse (although the composition of this plutonium is not suited for the production of nuclear weapons, see section Radioactive waste). The first and third steps both entail the risk of a potential deliberate misuse of nuclear material. Here the keyword is the *proliferation* of nuclear technology. We will treat each of these three steps in greater detail.

Enrichment: ultracentrifuges
Since natural uranium, as mined in uranium mines, contains only 0.7% of the fissile component U-235, it must be enriched before it can be used in the reactor. Chemical methods are ineffective, since the isotopes are chemically identical. However, use can be made of the difference in mass– however small – between U-235 and U-238. The enrichment process requires the components to be in a gaseous form. Therefore, the uranium is converted into uranium-hexafluoride (UF6), which is solid at an ambient temperature but turns into a gas at 65°C. Moreover, the important advantage of this process is that fluorine comes in only one single mass, as fluorine-19. Thus we avoid additional complications in the separation process.

Two approaches are currently in use in the process of separating the 'light' from the 'heavy' component. One is based on diffusion (the heavier UF_6 molecules diffuse somewhat slower that the lighter ones), while the other uses a centrifuge (with the heavier ones being more greatly influenced by the centrifugal forces than the lighter ones). This latter process was developed in the Netherlands, in cooperation with researchers from Germany and the UK. The enrichment process is performed by Urenco in the Netherlands, Germany, UK and USA.

Enrichment of uranium is no doubt a 'sensitive' technology. If the enrichment process is repeated often enough, it produces high-enough concentrations of U-235 so that the uranium can be used to produce a nuclear bomb. This so-called '*weapons-grade* uranium' requires that 80 to 90% of the nuclear material is U-235: the larger the concentration of U-235, the smaller the critical mass that is necessary to produce a bomb. The bomb that was dropped on Hiroshima on 6 August 1945 consisted of approximately 60 kg of uranium with a U-235 concentration of about 85%. Concentrations below 80% would require larger masses and become impractical.

The reactor itself

Nuclear reactors must be operated with extreme care, especially concerning safety issues. It is useful to know just where possible problems may arise.

The energy released during fission reactions is extremely large. This heat is removed via the 'primary cooling circuit' and – after a few intermediate steps – the heat is used to drive a steam turbine (see figure on page 143).

But suppose the cooling process fails for some reason. In this case, the chain reaction will automatically stop by itself: if the water used in the cooling process expands or evaporates because of rising temperatures, the neutrons will no longer be sufficiently slowed down to create a reaction. And, remember, the slowed-down neutrons are necessary to keep the reaction going.

But this is not the whole story yet. Even after the fission reaction process has been halted, a lot of heat will still be produced by the radioactive decay of the fragments. Immediately after the reaction has been halted for whatever reason, the residual heat production is 7% of the thermal power during normal operations. This is still a very large amount of heat being produced in a relatively small space. This 'residual heat' must be actively removed by means of a cooling system using pumps. Should this emergency process fail, the temperature will, in a few minutes, rise to a point where the fuel rods can be damaged. Eventually, this may lead to a (partial) meltdown of the reactor core. This is what happened in some of the reactors at the Fukushima Dai'ichi power plant in March 2011 in Japan, after an earthquake-induced tsunami caused the cooling systems to fail.

To avoid such disaster scenarios in the future, so-called 'inherently safe reactors' are being developed (see box page 149).

Why is there such an enormous residual heat production even after the reactor is shut down? The basic reason is the following. The parent nucleus, U-235, has many more neutrons than protons. This is necessary to keep the nucleus with all those repelling protons together. In case of uranium-235, there are 143 neutrons and only 92 protons (adding up to 235), which puts it in the extreme upper right corner of the graph on p. 146. After fission, the two fragments automatically have a similarly high neutron-to-proton ratio. For the sake of the argument let us assume that the uranium nucleus, after catching a neutron, breaks up in two equal halves, each having 46 protons and 72 neutrons. These smaller nuclei end up *above* the band of stable nuclei, because of the curvature of the graph. Their number of neutrons is much too large to be stable.

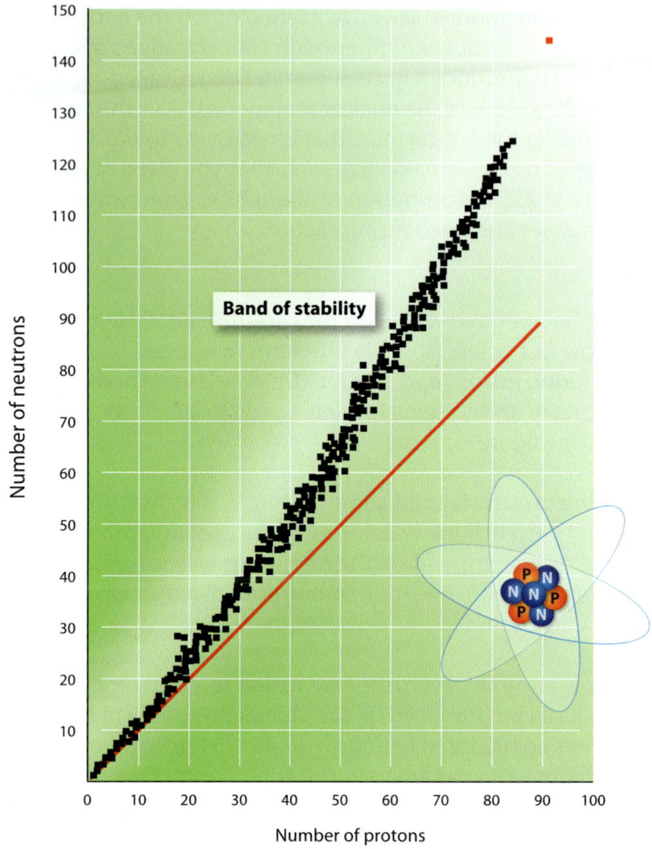

The number of neutrons as a function of the number of protons for stable nuclei, represented by black squares. For equal numbers of protons and neutrons, the nuclei would follow the red line. We see that light nuclei do have about the same number of protons and neutrons, whereas heavier nuclei have many more neutrons than protons. Uranium-235, like all nuclei having more than 83 protons, is not a stable nucleus, and is therefore not part of the black 'band of stability', but it is added as a red square in the upper right corner.

As a result, they seek stability, moving towards the band of stable nuclei by emitting particles. The emitted particles are primarily β-particles, *i.e.*, fast electrons. These originate from a neutron being converted into a proton plus an electron (which keeps the total charge at zero). So, each time a nucleus emits a β-particle, it looses a neutron and gains a proton. In the graph, it moves one step down *and* one step to the right, thus getting closer to the band of stability. In this cascading process, short-lived intermediate nuclei are formed. An example of such an intermediate nucleus is Iodine-131. It is also a β-particle emitter, and has a half-life of 8 days.

This example illustrates that the radioactive decay of the fission products is initially a fast process. Thus, the heat produced in the beginning of the process is also large. Gradually, as the intermediate fission products get closer to the band of stable nuclei, their activity decreases, and the residual heat production goes down accordingly.

Because of the residual heat production, used nuclear fuel rods are cooled in a water basin near the reactor for at least several months after being removed from the reactor. Such spent-fuel pools require active cooling. If this fails for some reason, the temperature will become high enough to cause serious damage to the fuel rods, and even to the building in which the pool is located. Radioactive material can be released in the process. This happened in some reactor buildings in the Fukushima Dai'ichi plant, although the reactors themselves had already been shut down for maintenance before the earthquake struck. This illustrates that not only the reactor itself, but also the spent-fuel pool must have guaranteed cooling under any circumstance.

After the heat release from the used fuel has gone down sufficiently to consider reprocessing or disposal, it can be removed from the power plant.

Radioactive waste

Spent fuel rods contain mainly depleted uranium, plutonium and fission products. How much of each is produced in a typical nuclear plant in a year?
Let us consider a power plant that produces an average of 1,000 MW of electricity (modern reactors can produce 1,500 MW). After one year of operation the used nuclear material of such a plant consists of 1,000 kg of fission products, about 200 kg of plutonium-239 and about 30,000 kg uranium.

There are two options in considering the next step.
The first option is to remove the plutonium and uranium from the used nuclear fuel so that these can be 'reprocessed', which is done in a reprocessing plant. This option may not always be commercially feasible, depending on the price of uranium.
The second option is to simply dispose of all of the used fuel as nuclear waste.

A few factors play a role in the decision-making process regarding the best option. This includes the key question of what to do with the plutonium produced by the reactor.

- If the plutonium is removed from the nuclear waste it can be re-used as nuclear fuel. This is especially important for breeder reactors (see section A4).

Incidentally, this plutonium mix is unsuited for the production of a nuclear bomb because the fissile plutonium-239 is contaminated by plutonium-240, which undergoes fission spontaneously and would ignite a bomb prematurely. It is also contaminated by plutonium-238, which produces a great deal of heat and requires continuous cooling. Bomb-grade plutonium contains mainly plutonium-239 and less than 7% plutonium-240. The atomic bomb dropped on Nagasaki contained about 7 kg of plutonium-239.

Other than plutonium, nuclear waste consists mainly of fission products with relatively short half-lives. These are primarily strontium-90 (half-life of 28 years) and caesium-137 (half-life of 30 years). For an explanation of 'half-life', see section C2, page 162. A 30-year half-life means that the waste must be stored for three centuries before the radioactivity has been reduced to 0.1% of its original value (this is because $(\frac{1}{2})^{10}$ is roughly 1/1,000).

• If the plutonium is *not* extracted from the used fuel, it has to be stored for an extraordinarily long period of time because the half-life of plutonium-239 is 24,000 years. On a geological timeline, however, this is relatively short, which is why plutonium-239, unlike uranium, does not occur in nature. This relatively 'short' time also means that it is far more radioactive than uranium. The radiation emitted consists of α-particles (with an energy of 5 MeV). These α-particles have a very shallow penetration depth because of their relatively large mass and charge, as explained in section C2. Therefore, plutonium-239 is safe as long as it is kept outside of the human body. But it is precisely this small penetration depth that makes it hazardous when inhaled or ingested, since the radiation's damage is limited to the very vulnerable lung and digestive tract tissue (see section C2).

We can thus conclude that if radioactive waste is *not* reprocessed (the 'once-through cycle') it has to be safely stored for about 200,000 years before the radioactivity has decayed to the level of the uranium from which it originated.

What are the choices that are currently being made to deal with nuclear waste, and what are the prospects?

The USA had by the 1970s already prohibited reprocessing operations, primarily because of the potential for misusing plutonium for destructive ends. In contrast, some other countries prefer reprocessing to other options. France, Belgium, Japan, Germany, Switzerland, the Netherlands and a few other countries transport their nuclear waste to a reprocessing plant in La Hague in France, where the plutonium is extracted from the waste. Subsequently, it has been re-used for several years now as nuclear fuel in the form of MOX, or Mixed Oxide. This is a mixture of oxides of plutonium and uranium, which is suitable as a nuclear fuel in most current reactors.
Finland is currently constructing a permanent nuclear waste storage facility in Olkiluoto, which should be operational by 2020.

There is an interesting example of an underground long-term storage that is found in nature, which gives some indication that safe long-term geological storage may be feasible. In 1972, natural 'fission reactors' were discovered in Gabon, which have survived at least several hundred thousand years, about 2 billion years ago. Its remnants (including the radioactive fission products and so-called transuranes) seem to have only moved a few centimetres in all that time, and have by now decayed down to harmless elements (see, *e.g.*, Wikipedia: *Natural Nuclear Fission Reactor*).

A3 | Present Role of Nuclear Fission World-wide

Nuclear reactors are primarily used for *base-load* electricity. This means that they operate almost continuously, which is partly for technical reasons and partly because nuclear fuel is cheap in relation to the investment costs. They are, therefore, not well-suited for responding to sudden changes in the demand for electricity (or changes in *supply*; just think of wind and solar power, or a power failure at a conventional plant). An example of fluctuations in electricity demand is shown in chapter 12 (page 167).

Inherently safe nuclear reactor (Source: J.L. Kloosterman, T.U. Delft)

How can we ignore a nuclear reactor's potential failures and accidents with 100% certainty? The nicest situation would obviously be to have a reactor that was 100% disaster proof even if all of the mechanical and electrical systems were to fail. This is called an *inherently safe reactor*. There is a design that may be able to meet this requirement – the *Pebble bed reactor*. It is sometimes called a 'generation 3+ reactor'. The concept is for nuclear fuel to be packed in small spheres of about 1 mm in diameter (see illustration). In turn, these small spheres are assembled into billiard-ball-sized pebbles each with a 6-cm diameter.

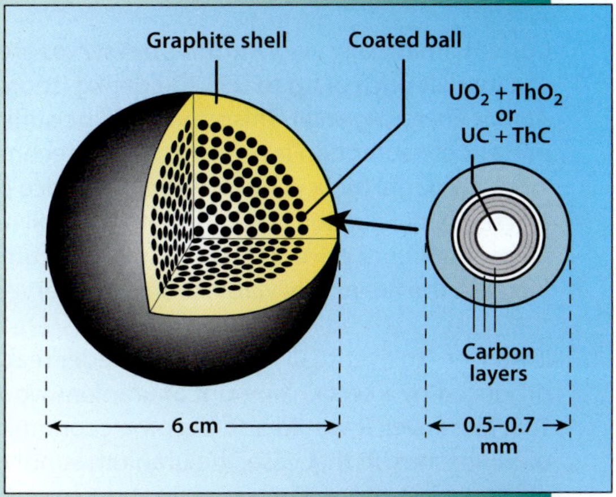

The pebbles are cooled by a strong flow of helium gas. This transports the heat to a turbine, so there is no water involved. Slowing down the neutrons to keep the reaction going is achieved primarily by the graphite in the pebbles and along the reactor wall, which is made of graphite. Graphite works fairly well as a 'moderator', since it consists of relatively light carbon atoms, which absorbs an appreciable part of a neutron's speed upon collision.

One of the features that make this type of reactor inherently safe is the fact that the chain reaction stops automatically when the temperature of the pebbles rises above a certain temperature (as a result of the Doppler effect). Granted, the temperature will rise because of residual heat production, but the pebbles are designed to withstand temperatures of up to 1,600°C. As long as the reactor is kept small, with a large surface-to-volume ratio, then the reactor core will never experience these types of high temperatures. An experimental reactor of this type located in Jülich, near Aachen, Germany, has already undergone about 20 years of testing during the 1980s and 1990s and passed all of its safety and security tests successfully.

Since these reactors must be kept small in view of safety, the power they generate is also relatively small: 160 MW of electricity maximum. Therefore a number of these reactors are clustered to larger units. A plant of this kind is under construction in China. This concept, called the High Temperature Reactor Pebble-bed Module (HTR-PM), contains two reactors of 250 MW thermal energy each connected to one steam generator producing 210 MW of electricity. These reactors should be in operation in 2015, but large-scale deployment is realistically foreseen for 2020 at the earliest.

World-wide, nuclear energy, generated by 440 nuclear power plants with a joint power of 400 GW electricity, contributes about 14% to the total world-wide electricity supply (more precisely, 13.5% in 2008; Source: International Energy Agency, IEA). This represents about 5% of *total* global energy demand.

Within the EU, nuclear energy's contribution is relatively high. In 2005, it contributed 30% of total EU electricity or 14% of the *total energy* needs (Source: IEA). *The Netherlands* has only one nuclear reactor that is used to produce electricity, the PWR reactor in Borssele. It has an installed power of 500 MW of electricity, and produces an average 450 MW. This represents 4% of total electricity produced, or almost 1% of *total energy* use.

Current, financially viable ***uranium reserves*** are estimated at about 4.5 billion (10^9) kg, at a market price of up to US$ 130 per kg (source: *World Energy Council* and *International Atomic Energy Agency*). This supply will be sufficient for the next 100 years, assuming present consumption levels and current technology (no breeding and predominantly 'once-through' fuel cycle). However, the price of uranium is only a small fraction of total electricity production costs. Therefore, it will remain economically viable to mine uranium at a higher price, even at double the current market price. Such a price rise would increase the financially viable uranium reserves by an order of magnitude.

If we were to eventually switch to breeder reactors (next section), the amount of energy produced by a certain amount of uranium would increase by a factor of approximately 100. Moreover, it would also become economically interesting to mine less lucrative uranium sites. In that case, the uranium supply could be expected to last for hundreds of thousands of years, given *present* consumption levels.

A4 | Breeder Reactors

The goal of breeder reactors is to utilise as much of the uranium's energy as possible by also using the non-fissile U-238. After all, this is by far the most abundant isotope, comprising 99.3% of the natural uranium. This is then converted into plutonium-239 in the reactor. This is performed by having the U-238 nucleus capture a neutron, as explained in section A1. But to achieve this, a sufficient number of neutrons must be available, in addition to the neutrons already needed to keep the chain reaction going.

However, it is found that the fission of a uranium or plutonium nucleus yields slightly more neutrons if it is hit by a *fast* rather than a slow neutron. This is why breeder reactors fuelled by uranium/plutonium use fast neutrons, and are therefore called *fast breeder reactors*. Using water as a 'cooling medium' to carry away the energy produced by fission is not practical in this case because it contains too many light nuclei (the H-atoms) that effectively slow the neutrons down too much. This is solved by using liquid sodium instead. Sodium consists of fairly heavy atoms, and a colliding neutron will bounce off a sodium atom like a pea hitting a billiard ball; it will change directions but will not be slowed down to any significant extent.

Fourth-generation nuclear reactors

(source: T. van der Hagen, T.U. Delft)

The European Union and ten other countries are currently engaging in research into six different possible 'fourth generation' nuclear reactors. Just like the 'inherently safe' *Pebble Bed Reactor* (see box on page 149) this type of reactor will not be in operation before 2020. The types include:

1. The *'Very High Temperature Reactor'* (VHTR):
This is a helium-cooled type of reactor with an exhaust temperature of about 1,000°C. It is suited for the large-scale, efficient production of hydrogen. Its basic operating principles have been demonstrated at the laboratory level in Japan.

2. The *'SuperCritical Water-cooled Reactor'* (SCWR):
This reactor is cooled using water at a temperature of around 550°C. As a consequence it could be very efficient in the generation of electricity.

3. The *'Gas-cooled Fast Reactor'* (GFR):
In this helium-cooled reactor the neutrons are *not* slowed down by water or graphite (which the word 'fast' already implies). Radioactive waste – like that of other reactors – can be destroyed and even recycled as part of its closed-circuit fuel cycle.

4. The *'Lead-cooled Fast Reactor'* (LFR):
This reactor is cooled by lead, so the neutrons are high-velocity neutrons. In other details it is similar to the GFR.

5. The *'Sodium-cooled Fast Reactor'* (SFR):
Is characterised by the fact that it is cooled by sodium, and the neutrons are high velocity. In other details it is similar to the GFR.

6. The *'Molten Salt Reactor'* (MSR)
This reactor has a liquid, circulating reactor core. It operates under low pressure (5 bar) and has a high exhaust temperature of at least 700°C. The focus is specifically on fast-neutron systems.

Fourth-generation nuclear reactors

Reactors using this process have been coined *Sodium-cooled fast breeder reactors*. So the word 'fast' here refers to the neutrons, not the reactor.

To achieve really efficient 'breeding' each fission reaction should produce a sufficient number of neutrons on average to keep the reaction chain active (which requires at least *one* neutron) plus at least one additional neutron that manages to convert a (non-fissile) U-238 nucleus into a (fissile) plutonium nucleus. This can best be achieved when plutonium-239 is used as nuclear fuel. When it undergoes the fission process it produces *three* fast neutrons, of which an average of slightly more than one produces a new plutonium nucleus. In this way, the reactor produces more fuel (*i.e.*, plutonium-239) than it consumes. Furthermore, it can reduce the amount of radioactive waste since some of the highly radioactive waste will undergo fission.

This type of breeder reactor was built during the 1970s in Scotland, Russia, Germany (near Kalkar) and in France (near Creys-Malville in the Rhône valley). The one in Russia is the only one still in use. The Kalkar reactor was never activated, mainly because of widespread social resistance, and it was eventually converted into an amusement park.

A5 | Pros and Cons of Nuclear Fission

An obvious advantage of nuclear power is that splitting uranium atoms does not produce CO_2. Furthermore, the construction of nuclear power plants uses very little fossil fuel. Moreover, nuclear power makes us less dependent on oil-producing countries. Roughly one-quarter of the total world uranium reserves are located in Australia. There is fuel for hundreds of years for the current active reactors. Creating a certain mix of energy sources makes the world less vulnerable to supply fluctuations. Another advantage is that operational costs are relatively low.

One drawback is that initial capital investment is very high at about 3 billion euros for a modern 1,500 MW 'third-generation' *European Pressurised Reactor* (EPR) and that the construction period is relatively long.
Another concern of the increased use of nuclear fission is that the proliferation of nuclear technology leads to the increased risk of misuse. Furthermore, there is no nuclear-waste storage method that has yet been proven to be safe in the long term. In addition, there is always a small chance of a catastrophic accident, as the Fukushima Dai'ichi plant has shown in 2011. Furthermore, nuclear power is not really sustainable as long as there is a dependence on the U-235 isotope. But, by switching to breeder reactors there will be an increased conversion of U-238 into plutonium, which must be carefully kept outside our body.

In short, the choice of nuclear fission energy or not is more of a moral and social issue than a scientific problem. In choosing between various options, one should keep in mind that any form of energy production has its drawbacks. There is no such thing as a free lunch. Somewhere, someone will have to pay the bill.

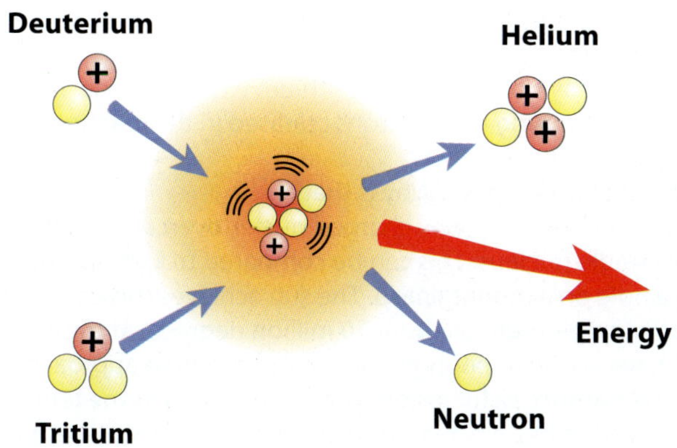

A schematic view of nuclear fusion. Shown is the so-called D-T reaction, in which a deuterium nucleus (D) and a tritium nucleus (T), both hydrogen isotopes, fuse to form a helium nucleus plus a neutron.

B | Fusion

Nuclear fusion is basically nothing new. The sun has been running on fusion energy for about 5 billion years now. The Greek name for the sun is Helios, from which 'helium' is derived, the reaction product of nuclear fusion. And as far as our understanding of the cosmos goes, nuclear fusion is its basic source of energy. An example of a fusion reaction is illustrated in the figure above.

It is a pity that, here on earth, we do not yet have an elegant process to harness nuclear fusion to produce energy, since it does show great promise for the future. This is especially obvious if we look at its potential.

B1 | The Potential of Nuclear Fusion

The biggest advantage of nuclear fusion is that it yields an energy supply that is unlimited for all practical purposes, as we will see below. One out of every 7,000 hydrogen nuclei in nature is, in fact, a 'heavy hydrogen' (a deuterium nucleus or simply D). If we were to convert all of the Ds in all of the ocean's waters into helium, we would be releasing an amount of energy equal to present global energy use over a period of tens of billions of years. Since our own sun is expected to keep shining for a few more billion years, there is enough reason to call the supply of fusion energy endless. Anyone interested can check this in the box 'Fusion reactions' on page 156.

B2 | Nuclear Fusion in Practice

Why is nuclear fusion so difficult to harness on earth? First, it should be pointed out that not all fusion reactions are equally difficult to achieve. The most attractive one is the D-D reaction (involving only deuterium nuclei), which is also the most difficult one to perform because it requires that reactants attain an enormous speed to facilitate fusion. The D-T reaction between a deuterium nucleus (consisting of a proton plus a neutron) and a tritium nucleus (a proton plus *two* neutrons) is less difficult. It is shown in the figure above.

The drawback, however, is that it involves tritium. Unlike the stable deuterium nucleus, tritium does not occur in nature because of its relatively short half-life of 12 years. So, it must be manufactured, from lithium, by bombarding the lithium with neutrons. This can be performed in the reactor itself by using the neutrons released in the reaction.

But even this D-T reaction is difficult to accomplish. Why is that?
Because we have to force the fusion of two positively charged nuclei, overcoming all of the repelling forces, to produce a single nucleus. This can be compared to kicking a golf ball up a steep slope to get it in a hole on a mountaintop. The sun achieves this constantly, but the process occurs at a temperature of about 10 million degrees. At these high temperatures the particles have acquired enough energy and speed to approach one another closely enough to achieve successful fusion. These kinds of extreme temperatures cannot be achieved on earth, since any material used would melt, way before we reach the desired temperature. So we have to use a trick.

One trick to achieve fusion involves so-called *inertial confinement*. The idea is that the materials to be fused – deuterium and tritium – are placed in small, hollow balls at low temperatures. These balls are then suddenly (within, say, a billionth of a second) heated to a hundred million degrees. This is done, for example, by shining powerful lasers or ion beams on the balls from all sides. The idea is to achieve fusion before the balls have a chance to disintegrate – hence the name '*inertial* confinement'. This method has thus far been primarily used for military applications.

The other trick has been the focus of international fusion research. It involves containing the nuclei by using magnetic fields, so-called *magnetic confinement fusion*. Why use magnetic fields? We should bear in mind that, at the high temperatures needed for fusion, all atoms and molecules in a gas are ionised, forming a so-called plasma. Since they have a net charge, they can thus be easily manipulated. For example, they can be deflected and guided along a curved path by means of a magnetic field, using the 'Lorentz force': any charged particle moving in a magnetic field has a force directed at it perpendicular to both its velocity and the magnetic field direction.

This creates an interesting effect because we can make charged particles like atomic nuclei spiral around the magnetic field lines, so that they are kept away from the walls of the containment vessel. This prevents the reactor walls from being hit by hot, energetic particles and from eroding or melting as a result. This can be seen in the figure below.

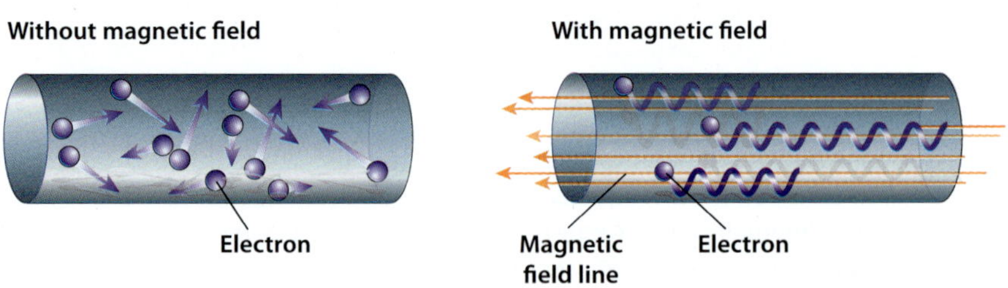

Confining charged particles using a magnetic field. The particles (in this case electrons) move forward in a straight line in a vacuum, as shown on the left. However, in the presence of a magnetic field, a particle's path spirals around a magnetic field line, which prevents it from hitting the wall. (Courtesy: N. Lopes Cardozo)

The 'container' is usually shaped like a gigantic doughnut, or *torus*, much like an inner tube of a car. It is often called a 'Tokamak', the name of the Russian container where this process was first tested. The magnetic field lines follow a circular path inside the *torus*. This way the charged particles (*i.e., all* of the particles except the neutrons) follow spiral-like paths around the magnetic field, never colliding with the Tokamak's walls except in a specially hardened region, the *divertor*, where the helium waste product is removed from the tokamak.

But this is still not sufficient to produce energy from fusion. The process has to deliver more energy than it uses. This is more easily achieved in larger facilities, where heat losses play a relatively smaller role, thus making larger fusion reactors more profitable. In fact, we actually have to move beyond a 'break-even point' to produce more energy than is used in the process. This is illustrated in the figure below. The *'Joint European Torus'* (JET) which was built in Culham, UK, was hovering around the break-even point, so the newer ITER (an acronym originally standing for International Tokamak Experimental Reactor) was introduced to produce ten times more energy than consumed. If ITER lives up to its expectations, it may lead the way and highlight the enormous progress being made in the field of nuclear fusion, which is appropriate, since ITER means 'the way' in Latin.

The role of a Tokamak fusion reactor's size is shown here. The power output, P_{out}, exceeds input P_{in} beyond the 'break-even point' (courtesy: N. Lopes Cardozo).

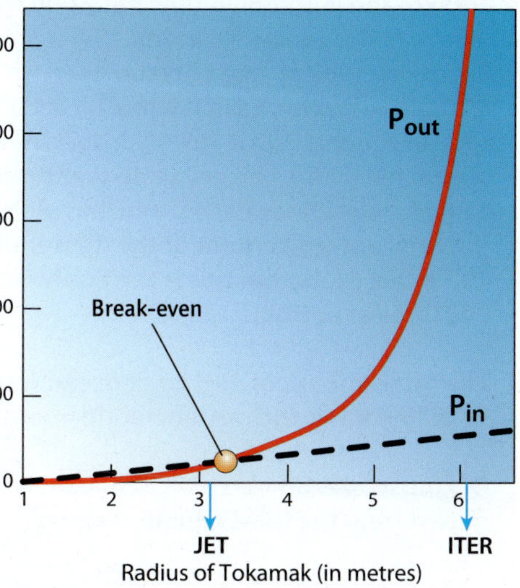

All of this implies that ITER, which is currently under construction near Cadarache in Southern France, will be large. The illustration on page 157 shows the size compared to an average human being (lower left corner). We should not be misled, however, by the size, since it is still basically smaller than a large coal-powered plant. Details about this project can be found, for example, at www.iter.org.
And once ITER is in operation in 2020 – hopefully – will we then finally be ready for fusion energy? Unfortunately, no. After the completion of ITER, about 20 more years of experimentation will be necessary to prove its scientific and technological viability. Meanwhile, demonstration reactors (DEMOs) will have to be built that will actually generate electricity. But construction on these is not expected to begin before 2025. Large-scale research projects will also be produced to sort out and test the most suitable materials to use in the construction of actual reactors. So, it is still quite some time before the first generation of commercial fusion reactors can even be constructed, let alone actually produce electricity. If everything goes according to plan, possibly 10 to 15% of total world energy production may come from nuclear fusion by the end of the century.

Fusion reactions

Let us calculate the potential of nuclear fusion for the world energy supply. To this end, we begin with the, as yet not technically achieved, D-D reaction, between two deuterium nuclei. One option is the following combination of reactions (the mass number, *i.e.*, the total number of protons plus neutrons, is the superscript on the left side of each nucleus):

$$^{2}D + {}^{2}D \rightarrow {}^{3}T + {}^{1}H + 4.0\ \text{MeV} \quad \text{followed by}$$
$$^{3}T + {}^{2}D \rightarrow {}^{4}He + n + 17.6\ \text{MeV} \quad \text{the net result being:}$$
$$3\,{}^{2}D \rightarrow {}^{4}He + n + {}^{1}H + 21.6\ \text{MeV}$$

In other words, this means that *one* deuterium nucleus yields a little more than 7 MeV. Since 1 MeV = 1.6×10^{-13} J we conclude that a single deuterium nucleus that fuses to form helium yields about 1×10^{-12} J.

We must find out how many deuterium nuclei we have available in the world's oceans. The total amount of sea water is estimated to be about 10^{21} kg, and it is known that 1 in 7,000 hydrogen atoms is a deuterium (heavy hydrogen or ^{2}D) atom.
Let us first look at 1 kg of water. It contains 1,000/18 moles or about 55 moles of water, each having $2 \times 6 \times 10^{23}$ hydrogen nuclei. This yields a total of roughly 700×10^{23} hydrogen nuclei. Since 1/7,000 of these are deuterium nuclei, we can calculate that there are 1×10^{22} deuterium nuclei in 1 kg of water. Since each D-nucleus yields more than 1×10^{-12} J (see above), the total energy content of the deuterium in 1 kg of water is more than 10^{10} joule. (Note that this is the equivalent to the energy contained in about 200 litres of petrol!)

The total amount of 10^{21} kg of ocean water thus produces 10^{31} J, or roughly 20 billion times the present world energy use per year!

Of course, we can also look at masses and use Einstein's famous $E = mc^2$ and depart from the third equation above.

The mass on the left is roughly $3 \times 2.0141 = 6.0423$ u.
The mass on the right is $4.0026 + 1.0087 + 1.0073 = 6.0186$ u.
The mass difference is 0.0237 u in 6 u, or 0.4%.

Let us look again at 1 kg of water. The hydrogen fraction in H_2O is 2/18 or 1/9 of which, again, 1/7,000 is deuterium. So, each kg of water contains a deuterium mass of 1/63,000 kg x 2 (since D has double the mass of H). This is 3×10^{-5} kg. In the comparison of masses above we concluded that 0.4% is converted into energy. To calculate the energy in 1 litre of water, we use $E = mc^2$:

$$E = 0.4\% \times 3 \times 10^{-5} \times 9 \times 10^{16}\ \text{J} = 1 \times 10^{10}\ \text{J}.$$

This corresponds nicely with the figures found above.

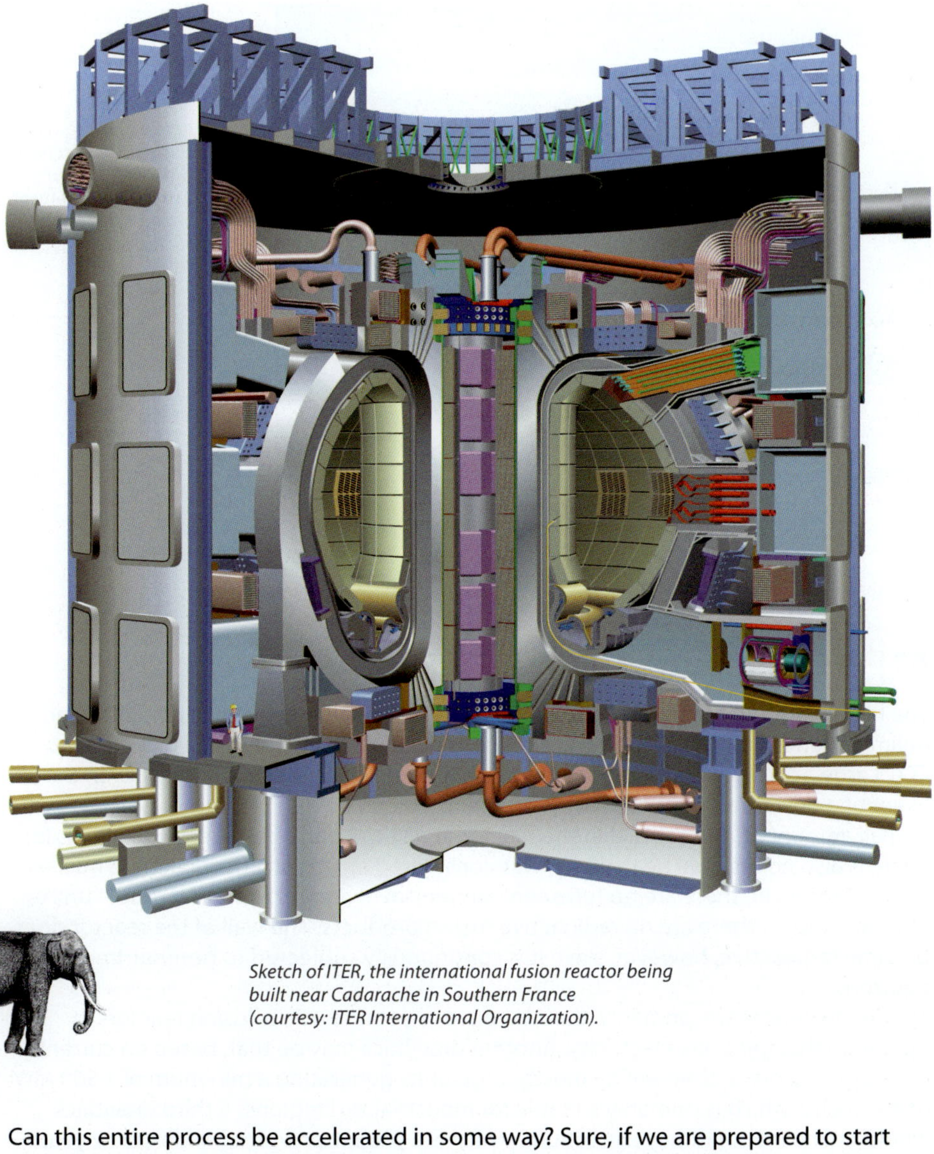

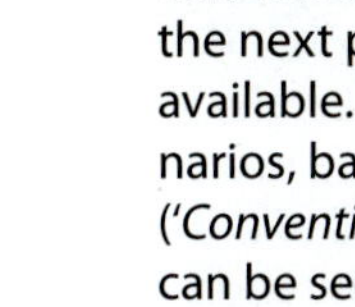

Sketch of ITER, the international fusion reactor being built near Cadarache in Southern France (courtesy: ITER International Organization).

Can this entire process be accelerated in some way? Sure, if we are prepared to start the next phase before all of the relevant data from the previous phase have become available. But even then it will not be really soon. The figure on p. 158 shows two scenarios, based on the official EU strategy: one that follows the usual development track ('*Conventional*') and one in which the various phases overlap somewhat ('*Fast track*'). It can be seen that the prediction, in any case, is that it will take at least 40 years, until the middle of this century, before we can expect any real contribution from nuclear fusion.

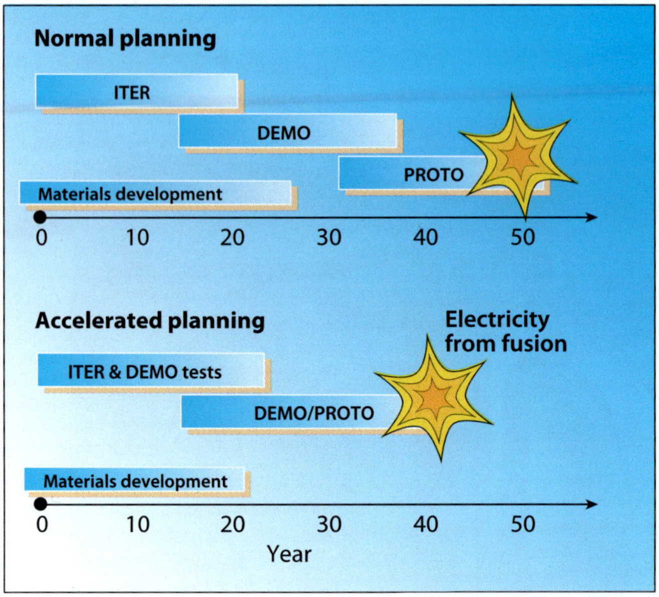

Two possible timeline scenarios for the development of nuclear fusion power. Upper panel: conventional development track. Lower panel: fast track with more overlap between development phases (courtesy: N. Lopes Cardozo).

B3 | Pros and Cons of Nuclear Fusion

The big advantage of nuclear fusion for energy production is that it is essentially an unlimited energy source. Nuclear fusion does not burn fossil fuels and produces no CO_2. Large accidents seem impossible, since the reaction stops automatically in the event of a power failure, which turns off the magnetic fields. Furthermore, there is only a small amount of nuclear fuel in the reactor at all times, and the energy density is low. There is also no residual heat, which was considered an important concern in nuclear fission. Moreover, there are no foreseen dangers involving proliferation. And – unlike nuclear fission – there are no radioactive fission products. The wall of the reactor does become radioactive, however, since it is continuously subjected to bombardment by neutrons.

The disadvantages are, primarily, that there are as yet no nuclear fusion reactors in operation that generate electricity. Another drawback may be that, based on current development plans, they will be mostly large units generating a minimum of 1,500 MW of electricity, which is primarily suitable for industrialized regions. A third drawback would be the fact that they use tritium, which is a β-emitter with a half-life of 12 years (see section Radioactivity and Health Hazards). These 'very heavy hydrogen' atoms, if embedded as hydrogen atoms into water molecules, pose a potential hazard. On the other hand, safety regulations limit the amount of tritium that can be stored at a fusion reactor at any time to a few kilograms. In the most dramatic case, where all of the tritium would leak into the atmosphere, dilution makes sure the tritium-concentration drops below health limits by the time it reaches the edge of the reactor site.

Finally, as mentioned above, there is the radioactivity of the reactor walls, which are continuously bombarded by neutrons. This means that various types of atoms making up the reactor wall will receive a surplus in neutrons, which then results in instability and thus radioactive radiation. They are, however, relatively short-lived, with a half-life of a decade, ensuring that a storage time of a century is enough to render the reactor vessel harmless.

The enormous potential of nuclear fusion means that it is worth pursuing research and development by all possible means. The nicest scenario would be that the research ultimately leads to simpler and smaller nuclear fusion facilities than those that seem to be presently possible.

C | Radioactivity and Health Hazards

C1 | Various Kinds of Radioactivity and Their Effects

Where does radioactivity come from? Radioactivity is the result of unstable nuclei seeking stability. They accomplish this by emitting alpha particles (α) and/or beta particles (β). They often emit gamma rays (γ) in the process. Such radioactive decay is completely spontaneous.

The resulting radiation is quite energetic, causing ionisation in human tissue, in other words, separating electrons from their molecules. It is therefore often referred to as 'ionising radiation', which can break the bonds between atoms in a molecule.

First, let's look at what these three kinds of radiation are and what their penetration depths are for various materials, including our own bodies.

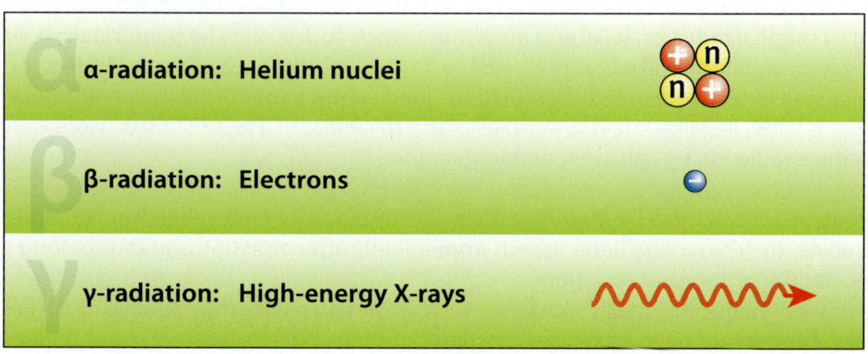

This illustration shows α-, β- and γ-radiation schematically

• The first are the **α-particles**, which consist of helium nuclei moving at relatively high speeds and having high energy. Therefore they can damage living cells. They are electrically charged since they contain two protons. This means that they interact strongly with anything they meet, so the depth of penetration is small. In fact, a sheet of paper or a few cm of air suffices to stop them. In other words, we can safely carry a chunk of uranium or plutonium around in our pockets without having to worry about any dangerous side effects.

However, we should prevent these α-particle emitters from entering our bodies. Since their penetration depth is so slight, they can easily damage the upper layers of vulnerable tissue in our lungs or digestive systems.

• The second kind are the **β-particles**, which consist of high-speed and high-energy electrons, somewhat similar to α-particles. However, they are very light compared to α-particles, since the mass of an electron is only about $1/2{,}000^{th}$ of the mass of a single proton or neutron. Given a certain amount of energy, they will therefore attain speeds that are much higher than α-particles. Consequently, they have a much shorter interaction time with anything they pass, so they move through objects much more readily. This means that their penetration is much deeper than that of α-particles. This is further exacerbated by the fact that their charge is only half that of an α-particle. As a result, they can penetrate a few metres of air, or a sheet of metal of a few millimetres. The penetration depth in our bodies is circa 5 mm, depending on the energy and type of tissue (see table on p. 161).

• The third kind are **γ-rays**, which are emitted along with the α- or β-particles. In fact, there is no pure γ-source; anything that emits γ-rays also emits α- or β-particles. Gamma-rays behave differently because they do not consist of particles but, like light, of electromagnetic radiation, but with a much shorter wavelength. They are best compared to X-rays. This means that the radiation does not have a specific penetration depth, like a bullet fired into a mound of sand. It works more gradually as it travels through a medium, like how light travels through coloured glass or dirty water. In other words, it decreases exponentially towards zero, like the temperature of a hot cup of coffee gradually decreases to the ambient temperature. We can characterise this attenuation in a certain medium by looking at the layer it takes to reduce the radiation to half its original value. After two such layers, the radiation will have been reduced to ¼ of its original radiation level and after three layers to 1/8, etc. The penetrating power of γ-rays is significant (like that of X-rays) so it takes a thick sheet of metal like lead or a thick wall of concrete to reduce radiation significantly. In our body the penetration power is large, just like X-rays. Examples of the thickness of a layer of tissue necessary to reduce radiation by half are found in the table.

• **Neutrons** should also be mentioned, since they play a key role in nuclear reactions. Moreover, they have very large speeds immediately after emission and thus belong to the category of 'ionising radiation'.
When these neutrons enter a substance such as human tissue, they behave much like γ-rays. They do not have a fixed penetration depth for a given substance, but their penetration behaviour is unpredictable because, after all, atoms consist mainly of empty space. The nucleus of an atom is like a cherry in the middle of a football stadium. Since neutrons are small and have no charge, there is a large chance that they will travel centimetres or metres through a substance without a single collision, but there is a small probability that they will collide with a nucleus along the way. The radiation intensity of these neutrons thus decreases gradually, like the γ-rays, to half its value after penetrating a certain depth of a given material. In human tissue, this distance is a few centimetres, depending on the neutron energy.

At this point, something unexpected happens. A beam of neutrons is attenuated by water or the human body more effectively than by a sheet of lead of the same thickness. This is due to the neutron's relatively small mass. If a light neutron collides with the much heavier lead atom, it changes course but not its speed. It's like the pea colliding with a billiard ball. In contrast, when a neutron penetrates human tissue or water, which consists mainly of hydrogen, it will hit mostly hydrogen nuclei (protons) with the same mass, which means that the exchange of energy is optimal. In a head-on collision, the proton bounces off at the initial speed of the neutron. But protons have a charge, so they strongly interact with matter. They ionise molecules along the way and their energy diminishes rapidly.

In turn, neutrons are usually captured by other nuclei, once they've expended their energy. In the human body this may very well be hydrogen nuclei, which are then converted into deuterium or 'heavy hydrogen' nuclei.

The penetrating powers of various types of ionising radiation when entering human tissue, in increasing order of penetration, in mm, and for two different energies, in MeV (see text). For the first three columns (α, proton and β), the distance refers to the penetration depth. For neutrons and γ-rays it is the distance travelled until the intensity is halved. Incidentally, the β-particles released by tritium during nuclear fusion have a much smaller amount of energy than listed here (0.01 MeV) and thus have smaller penetration depths (0.01 mm).

	α	proton	β	neutron	γ
Energy 1 MeV:	0.005	0.025	5	25	100 mm
Energy 10 MeV:	0.2	1.4	50	100	300 mm

The table is reasonably self-explanatory: charged particles (α, proton and β) have relatively short penetration depths, owing to their electric interactions with matter. The lightest among these has the deepest penetration depth. This is because, given a specific energy, lighter particles move faster and thus have less time to interact with atoms. Furthermore, it is reasonable to expect protons to travel not much further than α-particles, since their masses and charges are only slightly smaller. This is in stark contrast to electrons (β-particles), which are a few thousand times lighter, so they travel much faster, thus having less time to interact. Using the same 'interaction-time argument', we understand that higher energy means deeper penetration – hence the differences between the two rows.

Neutrons and γ-rays are not hindered by charges, so they travel a greater distance than the charged particles. At higher energy, a neutron beam may completely pass through a human body, losing half its intensity along the way. The other half is absorbed by tissue.

The energies quoted in the table should be compared with the energy necessary to ionise a molecule, or break a bond (a few eV, or electronvolt). This means that a particle with 10 MeV of energy can, in principle, affect a few million molecules. If the penetrating power is large enough, this effect will be scattered across a large volume. If it is small, the energy will be concentrated in a small piece of tissue and may do significant damage to that tissue.

The above scenarios assume that radiation intensity doesn't decrease if no matter is present to attenuate it. In other words, we have assumed that the 'beam' of radiation is a parallel beam, just like a sunbeam. This is automatically the case when the radiation source is far away. However, if the source is nearby, we have to take geometry into account. As we move away from the source, the intensity of the radiation decreases at the rate of the distance squared, since the total amount of radiation is spread across an area that *increases* with the square of the distance. This is similar to a nearby light shining on a book: if we move away from the light, the text becomes less visible. This is a purely geometrical effect and has nothing to do with attenuation by interaction with matter.

C2 | Half-life, Radioactivity, Doses and Hazards

Half-life and Radioactivity

How long does radioactivity last? A single nucleus' decay time is unpredictable: It may disintegrate immediately or over a period of time, thereby emitting 'radioactive radiation'. We can, however, calculate an average for many of these nuclei. The half-life of a radioactive substance is the amount of time it takes for it to lose half of its activity (and thus half its number of radioactive nuclei). A single half-life means activity has been reduced by half its original value. After two half-lives this is reduced to ¼, after three half-lives to 1/8 and so on. After 10 half-lives, the activity has been reduced by 2^{10}, or by a factor of 1,024, or roughly one thousandth of its original.

There is an intimate relation between half-life and radioactivity. If the half-life is long, the activity is smeared out over a long period of time, so the activity is low at any given moment. On the other hand, if the half-life is short, the activity is high.

Doses and Health Hazards

We should realise that nature already exposes us to a certain dose of radiation in the form of cosmic radiation, radiation from the soil and even from our own bodies (from potassium-40, for example). Thus its true hazards are dependent on its intensity and the type of radiation. For more insight into units, doses and the response of the human body, see the box 'Units and doses'.

The *natural dose* we are exposed to varies from location to location. In most countries this is 1 to 2 mSv (millisieverts; see box 'Units and doses') annually. It is emitted by radioactive substances in the soil and building materials, by radiation from space, and even by our own body (by potassium-40, circa 0.015 grams of which is found in each adult).

Units and doses

Radioactivity calculations may be confusing, with different quantities and their units. The three most relevant are:

Activity Unit: becquerel (Bq) = 1/s (one per second)
The activity of a radioactive substance is defined by the number of disintegrating nuclei per unit of time. This unit as defined by the international system of units is 'one per second', similar to one hertz (Hz) which is also 1/s. The difference between the two is that Hz is a rate at regular intervals, whereas Bq describes a *random* process, with an *average* per second. Therefore, it has its own name: becquerel (Bq) after Antoine Henri Becquerel, the French physicist who, in 1903, shared the Nobel Prize for discovering radioactivity, with the Curies.

Absorbed dose Unit: gray (Gy) = J/kg
The absorbed dose is defined as the amount of energy transferred to the human body by radiation, per unit of body mass. This is a well-defined and precise quantity, but it does not exactly measure radiation damage. It fails to take into account that some types of radiation are more hazardous than others, thus necessitating a more relevant unit:

Equivalent dose Unit: sievert (Sv) = J/kg
The equivalent dose is defined as the absorbed dose (in gray) multiplied by a 'quality factor', which depends on the type of radiation. It is 1 for β-particles, X-rays and γ-rays; 5 for protons; 20 for α-particles and 5 to 20 for neutrons, depending on their energy levels. Typical values for the *natural* dose that humans are exposed to are in the millisieverts range (see text).

Radiation from the soil varies greatly in different regions. Some locations may measure ten times the rate quoted above, because of radioactive elements like uranium and thorium.

Another source is *medical applications*. In Europe, this averages 1 mSv annually.

Radioactivity associated with *nuclear reactors* can play a role at several stages of the nuclear fuel cycle. We should distinguish here between radiation emitted by a source *outside* our body (primarily γ-radiation, being the only type having a long range), and radiation by a source *inside* our body. In the latter case α- and β-radiation play a dominant role. Because of the small penetration depth of these radiation types, α- and β-emitters are hazardous when inhaled or ingested, since the radiation's damage is limited to the vulnerable lung and digestive tract tissue.

Uranium, used as nuclear fuel, has low radioactivity because of its extremely long half-life (on the order of a billion years, see p. 142). It emits α-particles, which have a very small penetration depth, and is therefore relatively innocent as long as it is kept outside our body.

Plutonium-239, produced in a reactor from uranium-238, is also an α-particle emitter, like uranium. However, it has a much shorter half-life: 24,000 years, and consequently is much more radioactive. It should therefore be kept outside our body. According to the International Commission on Radiological Protection (ICRP), the safe threshold for inhalation is below one microgram per year. This is a very small quantity, although this is not unique to plutonium. For example, caesium-137 has a similar threshold of 2 micrograms.

Iodine-131 is an intermediate fission product (see p. 146). It is a β-particle emitter having a half-life of 8 days. In our body it is preferentially concentrated in the thyroid gland, where the primary health hazard is radiation-induced thyroid tumours. In the case when large amounts of iodine-131 are released as a result of a nuclear accident, as happened during the Fukushima disaster, children and pregnant women may be advised to take potassium iodide. This raises the level of 'normal' iodine (stable I-127) in the thyroid gland, thus decreasing the likelihood that the thyroid will absorb the radioactive species.

Caesium-137 and Strontium-90 are fission products having half-lives of about 30 years. Both are β-emitters. Inside the human body, caesium is more or less evenly distributed, and has a biological half-life of about 70 days. In contrast, strontium is a 'bone-seeker', because it is chemically similar to calcium. It is therefore primarily deposited in bone and bone marrow, where it can cause bone cancer.

D | Comparison of Various Ways to Generate Electricity

We have considered numerous energy sources, so it may be interesting to compare them. We can look at the requirements for the capacity to generate 1,000 MW of electricity on a continuous basis. This corresponds to the power of one large power plant, or roughly to the electricity consumption per one million people in a typical developed country (see table on page 65). The table on page 165 shows such a comparison.

Conclusions regarding Nuclear Power

- In its present form, nuclear energy is based on the ***fission*** of uranium, more specifically its least abundant isotope (*i.e.*, uranium-235). Nuclear energy provides 13.5% of the world's electricity production (2008), or about 5% of the *total* energy demand.

Various ways to generate 1,000 MW of electricity on a continuous basis. These are approximations. In cases requiring conversion from fuel to electricity a net efficiency of 1/3 has been assumed.

	Needed for 1,000 MW	CO_2 emission in kg/year
Coalf-fired plant	3×10^9 kg coal/year	10×10^9
Oil-fired plant	2×10^9 litres oil/year	8×10^9
Natural gas	3×10^9 m^3 gas/year	6×10^9
Nuclear power plant *	1×10^8 kg uranium ore/year, or 1×10^5 kg natural uranium, 1.5×10^4 kg enriched fuel, or 0.7×10^3 kg uranium-235	
Nuclear fusion plant	100 kg deuterium and 150 kg tritium/year	
Wind turbines (off-shore)	1,300 turbines with 90 m rotor diameter (600 km^2 of wind farm)	
Solar cells (15% efficiency)	60 km^2 of solar cell area	
Plants, trees (1% efficiency)	3,000 km^2 of land	
Human beings on exercise bicycles	10 million in continuous operation, or 30 million in 8-hour shifts	

* After a year in full operation, this yields approximately 1,000 kg of fission products and 200 kg of plutonium.

- Present uranium reserves can provide fuel for the current 440 nuclear reactors world-wide for about 100 years, assuming present nuclear technology and prices. This would be substantially longer if more expensive uranium supplies are included.
- If the most abundant uranium isotope (uranium-238) is also utilised, this time span increases by a factor of at least 100. This requires breeder reactors, however, which convert uranium-238 into plutonium.

- Investment costs of nuclear power plants are high (a few billion euros), but operating costs are low.
- Nuclear power plants are primarily suited for providing 'base load' electricity, not for responding to rapid fluctuations in demand (see chapter 12).
- The advantages of nuclear fission are that it makes the world less dependent on oil- and gas-producing countries, it is economically attractive and produces no CO_2.
- One disadvantage is that nuclear technology can be misused. Both uranium-235 and plutonium-239 (which is automatically produced in small quantities during operation) can be used to manufacture nuclear bombs. However, this requires a much more enriched form of uranium than used in reactors, or plutonium-239 extracted from used nuclear fuel.
- Although modern nuclear reactors meet every safety standard, serious accidents can occur in case of cooling system failure. This is mainly owing to the residual heat production after a reactor is shut down, as exemplified by the Fukushima disaster in 2011. 'Inherently safe reactors' are currently in the development phase and will not be available before 2020.
- Nuclear waste remains a problem that has not yet been solved. It concerns small quantities, though, which can possibly be stored safely in clay, granite or salt mines. If some of the more enduring components (like uranium and plutonium) are removed before storage, the waste decays relatively rapidly, reaching a rather innocent degree of radioactivity in a few centuries. If those components are *not* removed, this would take tens of thousands of years, because of plutonium's half-life of 24,000 years. Moreover, even tiny amounts of plutonium are hazardous, especially if inhaled.
- The drawbacks of nuclear power must be carefully weighed against the effects of fossil fuels on climate change.
- Nuclear ***fusion*** will not contribute to world energy demand before 2050. Fusion is still in the experimental phase. But if it becomes technically and economically feasible, its potential is enormous since fuel will last billions of years. One litre of ordinary water contains an amount of deuterium which produces the energy equivalent of 200 litres of oil.
- Fuel for fusion reactors is abundant and easily obtainable by everyone.
- Fusion reactors will probably be large units, comparable to current fission reactors that produce 1,500 MW of electricity.
- Problems related to radioactive waste will probably be much smaller for fusion reactors than for current fission reactors.
- Serious accidents seem improbable, and serious accidents affecting an area much larger than the plant itself seem impossible.
- Fusion research is currently being carried out in full openness by numerous nations. The experimental reactor ITER has a large number of cooperating countries: Japan, South Korea, India, China, the USA, Russia and the EU.

12 | Energy Storage

A | Why Store Energy?

Why would we want to store energy? First of all, because sometimes more energy is produced than can be used. An obvious example is solar energy for heating: save it in summertime when there is abundant sunshine and use it in winter when we need it most. Or wind power: store a surplus on windy days and save it for a windless day. Storage will be increasingly important as the contribution of sustainable energy increases.

Fluctuating energy *supply* aside, it is also handy to store energy to address fluctuations in *demand*. Electricity demand, for example, can fluctuate sharply. If production must match demand at all times this requires a large 'installed capacity' in terms of power plants. Capacity has to match even the single peak in demand occurring maybe once a year. This is expensive, since the rest of the year this over-capacity remains underutilised. Thus, it may be more economical to address these peaks by storing electricity. This is sometimes called *peak shaving*.

The figure below illustrates fluctuations in electricity demand. It reveals, for example, a steep rise in demand between 6:00 and 8:00 in the morning.

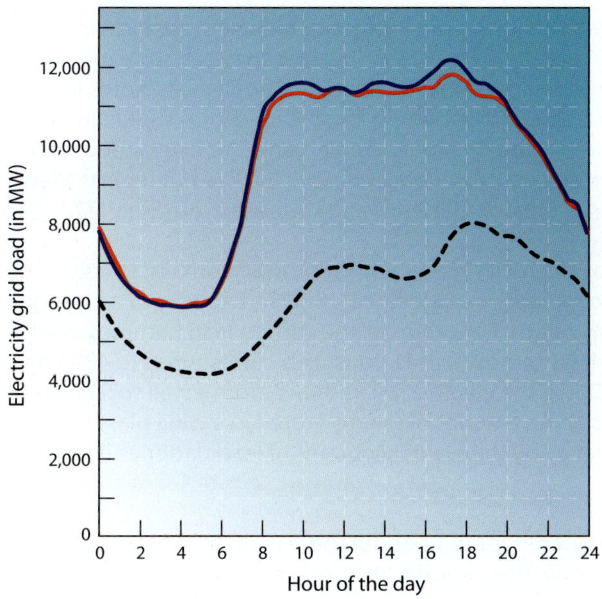

Example of fluctuating electricity demand over a 24-hour period. The upper curves show electricity demand on the busiest day (somewhere in December) in two consecutive years. The two are remarkably similar. The lower curve shows the demand on the least busy day during one year (in August). We see that demand varies between 4,000 and 12,000 MW during the year (courtesy EnergieNed, Arnhem, the Netherlands).

The figure also illustrates that there is always a certain base load, which is used throughout the year (4,000 MW, in this case). To meet this continuous demand, one usually employs the cheaper power plants that are, in turn, less adaptable to rapid fluctuations in demand. Examples include nuclear and coal-fired plants. However, it may be profitable to have these plants produce more than just the base load. They may then have a surplus at times, but this is no problem if the energy can be stored temporarily.

In addition to shifting energy use to a different *time*, we often want to have it in a different *place*. We are already doing this when we drive a car. In this case, the energy is usually stored as petrol or diesel, plus some in the battery.

By the time society makes the switch to large-scale use of sustainable energy like wind or solar, transporting energy will become an important issue. Just think of solar energy being transported from the Sahara to Europe in the form of hydrogen, for example. Or wind energy imported from a windy country.

B | Various Kinds of Storage

There are many storage options, ranging from dams and compressed air to batteries and flywheels. We will look at three important cases in more detail: storing electricity, storing heat, and hydrogen as an energy carrier.

B1 | Electricity

Storing electricity is not easy. Electricity in its pure form can only be stored in capacitors like those we use at home in our radios and TVs. However, they can store only small amounts of energy. Even the largest capacitor in our electrical appliances stores less energy than a fraction of a match does. Only modern super capacitors ('supercaps' or 'boostcaps'), used in experimental racing cars for example, can store large amounts of electricity. They can deliver a temporary power boost of up to 50 kW, but for just a few seconds. More details about this in the box 'Batteries and supercaps' on page 177.

The storage of larger amounts of electricity requires batteries. Incidentally, strictly speaking, batteries store *chemical* energy, not electricity. But the capacity of a battery is still modest compared to many needs. For example, the current lead batteries used in most cars contain about 40 Ah (amp-hours) at 12 volts. This corresponds to 480 Wh or 0.48 kWh of energy, about € 0.10 if the electricity was drawn from a wall socket. Put differently, this is the equivalent of the amount of energy stored in a wine glass full of petrol. The calculation shows that a few thousand kilograms of lead batteries is equivalent to 50 litres of petrol.

So, storing electricity in large quantities is not easy. But there is an alternative. Since mechanical energy can be efficiently (theoretically at 100%) converted into electrical energy, storage in the form of mechanical energy is an option. After all, artificial lakes have been used in this capacity for years. An example of a hydroelectric power station can be found near Vianden, Luxemburg (see photo).

The hydroelectric power station near Vianden, Luxemburg, is an example of storing electricity by using water as an intermediate.

If there is an electricity surplus, water is pumped into this higher-altitude lake, and when demand is high, the water is released to drive a turbine that generates electricity. The station delivers a maximum power of 1,100 MW at an overall efficiency of 74%.

To store substantial amounts of electricity we need a large amount of water and a reasonable difference in altitude. Creating an artificial lake in the sea, for example, is not an attractive option. Calculations show that it takes a 4-km-diameter lake and altitude differential of about 60 metres to generate 1,000 MW (typical power generated by a large plant) during a 20-hour period, with assumed efficiency of 75%. It may be more attractive to employ 'natural' height differences, such as in deserted coal mines. As the height difference increases, the energy per litre of pumped water also increases. Consequently, the amount of water needed is less than in the case of an artificial lake in the sea, for the same amount of stored energy.

There are other storage alternatives such as compressed air: Compressed Air Energy Storage (CAES). This may be performed underground, in deserted salt mines, for example. One drawback is that heat is released during the compression stage, and the same amount of heat must be supplied later on during expansion. The heat will reduce the efficiency of the plant unless it is temporarily stored. It is unclear whether this method will prove to be a viable alternative for hydroelectric power.

B2 | Heat

Anyone who uses a boiler to provide hot water already uses stored heat. But this is small compared to having to heat an average home during the winter. Thus we see that long-term heat storage for home heating purposes requires large amounts of water.

One would be tempted to think that water is not a good choice for storing heat but you would be wrong. In fact, water is a superior heat storage material. Its 'specific heat' is larger than those of metals or stone, for example. Only substances with a 'phase transition' (for example, from the liquid to the solid phases and vice versa) are better than water, provided the transition occurs at the right temperature. Think of the example of ice: we need to supply lots of heat to melt the ice, and the same amount of heat is released when it freezes again. All this takes place without changing its temperature: 0°C. So, such a phase transition would be a superior medium for heat storage, as long as the transition occurs at the right temperature.

A recent development using a transition is Thermo chemical Heat Storage. This process is currently being researched by the Energy research Centre of the Netherlands (ECN). It stores heat in the form of chemical energy by using a type of crystallised salt that absorbs water. Much heat is released during this absorption process. Since the process is reversible, the original situation can simply be restored by heating up the salt hydrate. Schematically this is represented as

$$AB + Heat \leftrightarrow A + B$$

where A and B are salt and water, respectively, and AB is the salt hydrate.

The volume of this system is only about 1/10 of the volume of water storing the same amount of energy. The nice thing is that we don't have to worry about insulating the system for long-term storage, since we are not storing heat (which would dissipate) but chemical components.

Even so, water may be a good heat-storage medium if we use it intelligently. Simply storing hot water in our own home to save the summer heat until winter is not clever. Suppose we want to store the equivalent of 1,000 m^3 of natural gas in the form of hot water, which would cool down by, say, 50°C while it heats the home during the winter. This would require a volume of 175 m^3 of water (the volume of a swimming pool), even if heat losses during the period between summer and winter were zero, which would be difficult to achieve. So we have to improve on that.

First of all, we need a *very large* volume, since the heat losses occur on the outer surfaces (proportional to the *square* of the dimensions) while the energy stored is determined by the volume (proportional to the *cube* of the dimension). Insulation is less critical here. This means that this system works better on a larger scale, to supply heat to a block of houses rather than a single home.

We can also employ heat pumps. These enable to the use of heat at a relatively low temperature, by 'pumping' it up to a higher temperature. As discussed in chapter 4, this can be done efficiently provided the temperature difference is not too large.

Drawing heat from a relatively warm reservoir (at, say, 10°C) and delivering it at, say, 35°C can be done efficiently. Temperatures of 35°C may suffice for floor heating, for instance. The heat could be drawn from underground water reservoirs, which are often supplied by nature (*'Aquifers'*, or natural groundwater storage areas). This is also related to geothermal energy (see chapter 10).

B3 | Hydrogen as an Energy Carrier

Hydrogen is not an energy source. It is an energy *carrier* because it is produced using other energy sources and occurs infrequently in nature. But once produced, it can be easily transported via pipelines like natural gas, or as a liquid.

Hydrogen is often written as H_2, since a hydrogen molecule consists of two atoms, like nitrogen. For hydrogen's relevant properties, see the table below.

Properties of molecular hydrogen (H_2)

Heat of combustion (higher*)	12.8 MJ/m³	or	142 MJ/kg
Heat of combustion (lower*)	10.8 MJ/m³	or	120 MJ/kg
Density (at 0°C, 1 bar)	0.090 kg/m³		
Boiling point	–253 °C	or	20.4 K
Density of liquid H_2	71 kg/m³		

* This is similar to natural gas. The 'higher' heating value includes the heat of condensation of the water vapour; the 'lower' value does not.

Producing hydrogen

There are several methods.

- One is the 'old-fashioned' way, using natural gas, oil or coal. Indeed, there are several hundred coal gasification plants world-wide. These plants first oxidise pulverised coal at high temperatures using oxygen and steam. The end products are H_2 and CO_2. If the CO_2 produced is captured and stored, this seems a sensible way of using the coal.

However, sooner or later we will run out of coal and other fossil fuels, so, for the sustainable production of hydrogen we need alternative methods, such as:

- By using electricity. This method is called 'electrolysis', because it splits water into hydrogen and oxygen using electricity as shown schematically by the upper arrow here below:

$$2\,H_2O \underset{\text{combustion or fuel cell}}{\overset{\text{electrolysis}}{\rightleftarrows}} 2\,H_2 + O_2$$

Modern equipment has an efficiency rate of 80%, meaning that 80% of the electricity consumed in the process is converted into energy in the form of hydrogen.
- Even nicer is producing hydrogen directly from solar energy, by processes called 'Photo-electrolysis' or 'Biophotolysis' (*cf.* chapter 8). But these techniques are as yet undeveloped. Photo-electrolysis has achieved an efficiency of about 10% in laboratory conditions.

Whatever method is used to produce H_2, the end product can be used at any time and location as a fuel. But 'fuel' should not be taken too literally, because there are more ways of using hydrogen.

Using hydrogen

Water is always the end product, as the lower arrow of the reaction on page 171 shows. There is no CO_2 involved. If the two processes above are carried out one after the other, we're back where we started, and haven't gained anything. In fact, we lose a bit, because there are always losses involved in conversion processes, which reduces overall efficiency. Hydrogen's efficiency as a fuel depends on how it's used.

- Simply burning hydrogen like natural gas is one option, of course. If it is used for heating, efficiency can be nearly 100%. Using it as a fuel for co-generation (electricity plus heat) may be a more economically attractive option.
- Hydrogen may become useful as fuel for internal combustion engines, like we currently use petrol. However, efficiency rates will not exceed 20 to 30%, which is the case for any internal combustion engine. If we want to use hydrogen to produce mechanical energy, we can do much better by using:
- A fuel cell, which works like electrolysis in reverse (see lower panel of the figure). A fuel cell directly converts H_2 and O_2 into electricity. Since an electric motor is more efficient than an internal combustion engine, this seems to have a future because we avoid the inefficient step of converting heat to mechanical energy. An efficiency of 60% in fuel cell operation has thus far been achieved, which is twice as efficient as the internal combustion engine.

So, a fuel cell offers a useful solution when we need electricity (or mechanical energy, since the two can be efficiently converted into one another).

Transport of hydrogen

Hydrogen is a clean fuel, which is especially useful for mobile applications like cars in the city. But it needs to be transported to where it's needed, and must be conveniently stored. Transporting hydrogen between two fixed points is easy. Hydrogen gas at a pressure of 20 bar, for example, can be transported through pipelines. This is economically viable for distances up to 1,000 km. Significant experience is available for this mode of transport, both in Europe and in the US. If distances become too large, another option is transport as a liquid. Since the boiling point is minus 253°C, this is not as easy as transporting oil.

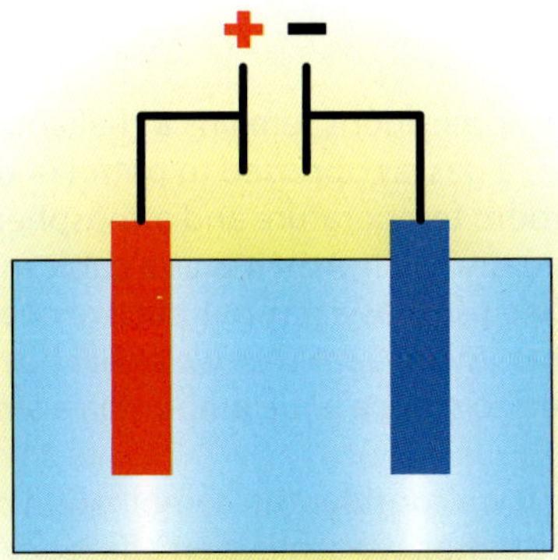

Anode — **Cathode**

$2\,H_2O \rightarrow 4\,H^+ + O_2 + 4\,e^-$ $\qquad$ $4\,H^+ + 4\,e^- \rightarrow 2\,H_2$

Net: $2\,H_2O \rightarrow 2\,H_2 + O_2$

Upper panel: production of hydrogen using electrolysis: Water plus electrical energy yields hydrogen plus oxygen.
Lower panel: use of hydrogen in a fuel cell. Hydrogen (upper left corner) and air (upper right) are fed into the cell. Hydrogen combines with the oxygen from the air, yielding water plus electricity. Note that one process equals the other in reverse.

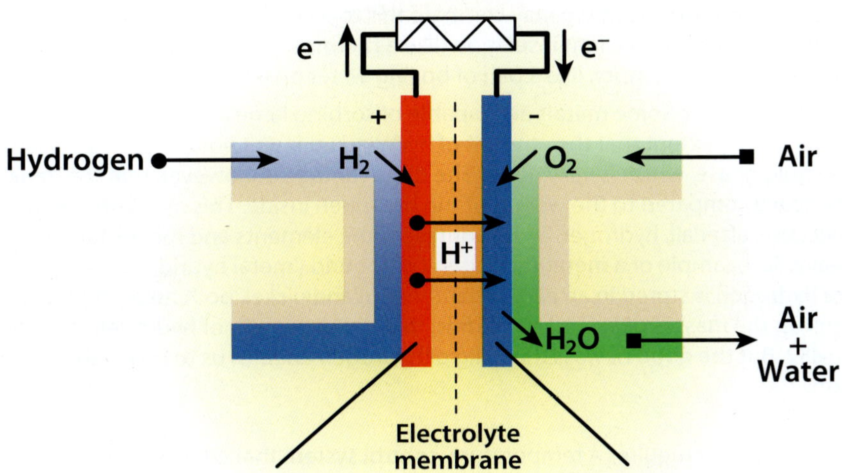

Anode — **Cathode**

$H_2 \rightarrow 2\,H^+ + 2\,e^-$ $\qquad$ $O_2 + 4\,H^+ + 4\,e^- \rightarrow 2\,H_2O$

Net: $2\,H_2 + O_2 \rightarrow 2\,H_2O$

Storing hydrogen in a car

Storing hydrogen for mobile applications remains a challenge. This is because, at ambient temperature, hydrogen is a gas, as opposed to petrol or diesel, which are liquids. When hydrogen is a gas at room temperature and atmospheric pressure, the number of molecules in a cubic metre is smaller than it is as a cubic metre of liquid by a factor of roughly one thousand. If we use it as a gas at 1 bar, we need volumes on the order of cubic metres rather than litres. This means that we either have to compress the gas or turn it into a liquid, if we want to store a large amount in a small volume.

We are lucky in one respect. If we consider the combination of hydrogen plus fuel cells as a way to power an electric motor, we achieve an efficiency twice that of an ordinary internal combustion engine. So, to drive a certain distance, we need to only store half the energy in the form of hydrogen.

There are currently three ways to store hydrogen in a car in a relatively compact volume.

1. **As a gas at high pressure** This requires strong (and thus heavy) containers. To store the energy equivalent of 50 litres of petrol, storage cylinders have a total mass of 300 to 400 kg. Incidentally, the fuel itself has a mass of less than 10 kg in this case. For city buses, for example, this is not really a problem. They have more space, weight is less crucial and they return to the garage frequently.
2. **As a liquid** As mentioned above, hydrogen only turns into a liquid at –253°C. This makes handling, loading and transport cumbersome. Moreover, the tank must be superbly insulated to prevent rapid evaporation. Even super-insulation means a 'boil-off' of 1 to 2% per day. We may compare this to a bottle of water in a 300°C environment. Even with excellent insulation (like a thermos flask) it will inevitably boil down and eventually empty, like a pan of boiling water on a tiny gas burner.
3. **Sorbed into a solid** Some metals are capable of sorbing large amounts of hydrogen, even to the extent that the amount of hydrogen per unit volume is larger than in the liquid phase, which is very good. The disadvantage is, however, that the 'tank' is very heavy compared to the weight of the hydrogen inside. This may seem obvious because, after all, hydrogen is the lightest of the elements and metals tend to be heavy. An example of a metal-hydrogen compound ('metal hydride') is $LaNi_5H_6$, where hydrogen is stored in an alloy of lanthanum and nickel, both heavy metals. If we look up the masses of La and Ni, we find that the mass ratio of hydrogen to metal is 6 to 432. But the good news is that 6 kg of hydrogen enables us to travel about 600 km.

Using metal hydrides requires a temperature-control system that controls sorption and desorption, since this process is very temperature dependent. It's comparable to a fluid's evaporation and condensation process.

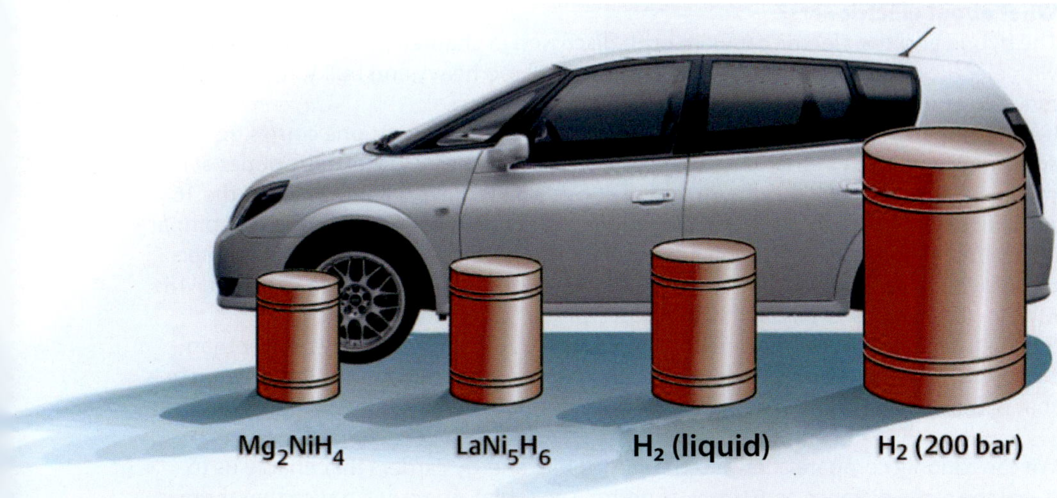

Artist's impression of the volume of hydrogen needed to drive 400 km, for various storage systems. From left to right: two metal hydrides; liquid hydrogen; and hydrogen gas at a pressure of 200 bar. The last tank could be substantially smaller if higher pressures are used (350 to 700 bar).

Given these complications, we can predict that hydrogen vehicles will, at least in the short term, fill their tanks with hydrogen gas at high pressure (350 to 700 bar). The 700 bar is already being used in Norway, on the HyNor, see below. So far, carrying liquid hydrogen on board is only practical in exceptional cases, like for racing cars. This is still too complicated for ordinary cars. Storage of H_2 as a hydride in metals on a large scale will, alas, require more research and development.

The figure above shows the 'tank' volume of various options, with the driving range of a tank of hydrogen fuel estimated at 400 km (4 kg of H_2).

Hydrogen on the 'Hyway'?

Obviously, re-fuelling with hydrogen along the highway requires a special infrastructure: 'gas stations' that deliver hydrogen, with highways that provide hydrogen refuelling called 'hyways' for obvious reasons. California began a hydrogen highway network in 2005, primarily to reduce air pollution, aiming to have 50 to 100 operable refuelling stations along the West Coast by 2010. There were only about 30 such filling stations in 2010, with several more under construction. Future development depends on how many hydrogen cars are sold, and vice versa, which implies slow growth. Europe has also several of these projects. For example, Germany, Sweden and Denmark have plans for a system of interconnected hyways. Norway, meanwhile, is building a 580-km hyway between Oslo and Stavanger (HyNor, or Hydrogen road of Norway).

There is one drawback: leaked hydrogen enters the atmosphere, where it may affect the ozone layer. The severity of this has yet to be determined, but there is reason enough (also in view of security) to insist on leak-proof refuelling stations.

What about electric cars?
The hybrid and the electric car were both discussed in chapter 5. The basic problems with electric cars are twofold: the batteries are still too heavy and bulky, reducing the range to about 150 km, and 'refuelling' still takes too long.
If one compares the refuelling rates of petrol *vs.* electrical energy, one comes up with some amazing figures. Charging a car battery at a standard 230 V electric outlet at home may be done at a maximum rate of about 3.5 kW, if we want to prevent the 16-Amp fuse from blowing. Just how slow this transfer rate is becomes clear if we compare it to petrol. Refuelling a normal car at a standard service station takes about 100 seconds for 50 litres. The transfer rate is thus 0.5 litres per second or about 18 MJ/s, or 18 MW! In other words, this is 5,000 times faster.
So, the future electric car may need services such as battery exchanges rather than battery recharging. This reminds us of the old days of the stagecoach. The difference is that in those days *horses* were exchanged, not batteries.

We can roughly compare a few alternative mobile energy carriers that enable us to travel the same distance as 50 litres of petrol (see table). Again, we notice that there is no energy carrier as convenient as petrol or diesel. The point is that nothing can beat the energy density of a fuel that is liquid at ambient temperatures. This basically limits us to hydrocarbons having about 10 carbon atoms per molecule, like oil, petrol and diesel. Fuels with shorter carbon chains like propane have vapour pressures that are too high to be very useful. Longer-chain molecules, such as fats, have vapour pressures that are too low. Alcohol may be one of the only alternatives. But alcohol production itself requires energy.

This may add an extra incentive for *conserving* oil. It is too handy a fuel to simply burn like scrap wood. The alternatives are much less convenient for transportation.

A rough comparison of various mobile energy carriers with an energy equivalent of 50 litres of petrol

	Volume in litres ('tank' included)	Mass in kg ('tank' included)
Petrol	52	60
Hydrogen under high pressure	300	400
Hydrogen, liquid	100	150
Hydrogen as $LaNi_5H_6$	200	500
High-tech flywheel	400	1,000
Lead batteries	2,000	3,000

Batteries and supercapacitors

Storing electrical energy is not easy. Almost all of the current model cars still rely on traditional lead batteries, but they don't have an awful lot of energy. An average 12-volt battery has a charge of 40 Ah, so it can deliver a 40 Amp current for one hour. Combining the 2 figures yields 480 Wh, or 0.480 kWh of energy. This is the equivalent of a wine glass of petrol. Taken per kg, this battery contains about 0.03 kWh/kg of energy, which is not much.

But we shouldn't despair. The nickel-metal-hydride (NiMH) battery, for example, does a much better job. It is relatively cheap, safe and contains 0.06 kWh/kg of energy, which is two times better than the lead battery. It is currently used in hybrid cars.

Lithium-ion batteries in mobile phones and laptops perform even better, achieving 0.14 kWh/kg. However, they are relatively expensive, complex and present some safety issues if used in cars. Moreover, the Li-ion battery is not particularly durable as a result of various chemical processes. It looses performance more quickly at higher temperatures and when fully charged. Optimum storage conditions imply lower temperatures with the battery loaded to 50% of capacity. Useful information is found at: www.batteryuniversity.com.

Batteries have problems delivering large amounts of energy quickly, *i.e.*, if they must produce a high power. This is because the chemical processes that form the basis of the storage mechanism react slowly. True electricity storage occurs in capacitors, which do not have these problems. They can be charged and discharged extremely quickly. But common capacitors such as those found in electric appliances are totally useless in storing large amounts of energy; they only contain a small charge, so they are exhausted very quickly when supplying high power. Modern 'supercaps' have a much larger capacity. They are extremely well suited for storing the braking energy of a car or a trolleybus, and subsequently delivering it back for quick acceleration. They are also called 'boostcaps' since they give brief boosts to racing cars. Another advantage is that they can be charged and discharged millions of times without performance loss, in contrast to batteries with a much more limited number of charge/discharge cycles. One drawback, however, is that, while they deliver large amounts of *power,* they cannot store large amounts of *energy*. Also, they are still expensive. Here is a list of approximate energy densities of batteries and supercapacitors, in kWh per kg:

Lead battery: (40 Ah $\times$ 12 V $\approx$ 0,5 kWh)	0.03	kWh/kg
NiMH battery	0.06	
Li-ion battery	0.15	
Li-ion polymer battery (LiPo)	0.20	
Supercapacitor	0.005	
predicted	0.020	kWh/kg; power 2 kW/kg

Conclusions regarding Energy Storage

- Storing *heat* in plain water is relatively efficient and easy for short-term storage (a few days). A hot water 'solar boiler' is one example.
- Using water for long-term heat storage, for example for six months of home heating, is only feasible in larger units. One way to accomplish this is to use underground reservoirs, found in nature in the form of 'aquifers'. Heat pumps can further boost efficiency.
- New developments such as thermo-chemical heat storage may reduce the required volumes substantially and are especially useful for long-term storage.
- Large-scale storage of *electricity* is most efficient in the form of mechanical energy, as found in high-lying reservoirs or lakes, for example.
- Small-scale electricity storage, *e.g.*, for electric cars or laptops, can be done with modern batteries, like metal hydride or lithium batteries. Although these are superior to current lead batteries, they are heavy and cumbersome compared to fossil fuels for mobile applications.
- Super capacitors are an option in situations requiring the quick storage and re-use of electrical energy, since they can handle high powers. Applications include the storage and re-use of braking energy.
- Hydrogen as an energy carrier promises a bright future. It can be produced from water by applying electrical energy. In turn, hydrogen can produce electricity directly in fuel cells. The closed cycle has an efficiency of circa 50%. The production of hydrogen from solar energy is currently being researched.
- One drawback of hydrogen for mobile applications is that storage remains a problem. The least difficult option is gaseous hydrogen under high pressure, about 350 to 700 bar. Hydrogen refuelling stations are currently under construction along highways in several countries.
- For ordinary vehicles, nothing is handier than a fuel tank filled with petrol or diesel. All of other options are heavier, bulkier and more cumbersome.

Epilogue | How to proceed?

We live in an exceptional era: the era of convenient energy. It is something we may not always realise. When we refuel our car, for example, we take energy on board at a rate which is five thousand times faster than if we would charge our electric-car batteries from the wall socket in our home. And the fuel we use has only a tiny fraction of the weight that it would have if we would drive on batteries. Air travel would not be possible at all without the use of a liquid fuel. In our home, energy enters through wires and pipes, silent and invisible.
The fact that energy is so convenient (and cheap!) has lured us into a rate of energy use that is not sustainable for the generations to come. The main reason is that we are simply running out of fossil fuel. A second reason is that we may be putting our climate into jeopardy.

What shall we do?
For many of us – those living in the affluent societies of the globe – the first remedy may be easy: have a critical look at our life style – fly less, use carpooling, drive more slowly, heat our homes more efficiently. In short, use energy wisely, and be aware of its link to justice and world peace.
For *all* of us, inhabitants of the globe in this unique era, there is an additional challenge: develop alternatives. Put a lot of research and development into cheaper and more efficient solar cells. Find ways to improve photosynthesis. Explore the possibilities of geothermal energy. Develop smart energy storage systems.

Since the nuclear disaster in Fukushima in March 2011 we have come to realise that the world is at a crossroad. Will nuclear fission plants be phased out (as, for example, Japan, Germany and Switzerland are planning), and will the world opt for a nuclear-fission-free future? Or will nuclear fission become a major energy source in the years to come?

In the latter case nuclear energy, which generates 13.5% of the world's electricity or about 5% of *total* energy, will have to be expanded substantially to play a major role. This will necessitate a switch to breeder reactors in view of finite uranium reserves.
In the former case, we need to buy time: time to let real sustainable energy (including nuclear fusion) come of age. A consequence may be that world energy use will not be able to continue to grow along the 'business-as-usual' scenario. Smarter and more sensible ways of using energy will then be required. Our way of life with its abundant energy use may have to be adjusted. In this perspective, the recession that affects the world since 2008 may, in hindsight, be considered a blessing in disguise.

Whatever choice will be made, it is imperative that we act, in all possible fields. What we do (or do *not* do) will largely determine the life of future generations. And it will determine the way future generations will look back upon us. Let's give them a future that we can be proud of.

Jo Hermans
Leiden, November 2011

Acknowledgement

It is a pleasure to thank colleagues, students and friends who have contributed to make this book better and my work lighter:
Jelle Brill, Robin Purchase, Roel Smit and Huub de Groot (Leiden University); Amanda Coady; Tim van der Hagen and Jan Leen Kloosterman (Delft University of Technology); Gareth Jones (Imperial College, London); Lauran van Oers (CML, Leiden University); Wim Sinke (ECN); Gieljan de Vries (FOM Institute Rijnhuizen).
In particular, I am very much indebted to the students of Leiden University College The Hague for their most valuable comments, especially to Maxime Verbeij, Alinta Geling, Kristel Hoogstad, Fabian Kemps Verhage, Ruben Dunsbergen and Lucas Antonisse.
Last but not least I want to thank my wife Hanneke for her endless patience and support.

Jo Hermans
Leiden, November 2011

Useful websites

www.iea.org	Extensive data, also for separate countries (International Energy Agency)
www.energy.gov	US data about Energy and Environment (US Department Of Energy)
www.eere.energy.gov	Energy Efficiency and Renewable Energy Programme (US) (US Department Of Energy)
www.reuk.co.uk	Much info about alternative energy from the UK (Renewable Energy UK)
www.worldenergy.org/publications	Data also about energy prices and renewable energy (World Energy Council)
www.bp.com	Various energy data; see, *e.g.*, Statistical Review of World Energy (BP Energy group)
data.worldbank.org/topic/energy-and-mining	Data about energy and mining; indicators; data catalogue (The World Bank)
www.nrg.eu	Information about nuclear energy, radiation, radioactive waste (Nuclear Research and consultancy Group)
www.iter.org	Everything about nuclear fusion and ITER (ITER Fusion international project)
www.ecn.nl	Various energy data about the Netherlands (Energy research Centre of the Netherlands)
re.jrc.ec.europa.eu/pvgis/solres/solreseurope.htm	Extensive data on solar radiation (Joint Research Centre of the European Commission)
setis.ec.europa.eu	Information system on key low-carbon energy technologies (Strategic Energy Technologies Information System, European Commission)

Index